Sets and integration
An outline of the development

Sets and integration
An outline of the development

D. van Dalen
Lecturer at the University of Utrecht

and

A. F. Monna
Professor at the University of Utrecht

WOLTERS-NOORDHOFF PUBLISHING   GRONINGEN
THE NETHERLANDS

ISBN-13: 978-94-010-2720-5   e-ISBN-13: 978-94-010-2718-2
DOI: 10.1007/ 978-94-010-2718-2

Library of Congress Catalog Card Number: 71-184993

Contents

Foreword

D. van Dalen   *Set theory from Cantor to Cohen*

A. F. Monna   *The integral from Riemann to Bourbaki*

Foreword

The present text resulted from lectures given by the authors at the Rijks Universiteit at Utrecht. These lectures were part of a series on 'History of Contemporary Mathematics'.

The need for such an enterprise was generally felt, since the curriculum at many universities is designed to suit an efficient treatment of advanced subjects rather than to reflect the development of notions and techniques. As it is very likely that this trend will continue, we decided to offer lectures of a less technical nature to provide students and interested listeners with a survey of the history of topics in our present-day mathematics. We consider it very useful for a mathematician to have an acquaintance with the history of the development of his subject, especially in the nineteenth century where the germs of many of modern disciplines can be found. Our attention has therefore been mainly directed to relatively young developments.

In the lectures we tried to stay clear of both oversimplification and extreme technicality. The result is a text, that should not cause difficulties to a reader with a working knowledge of mathematics. The developments sketched in this book are fundamental for many areas in mathematics and the notions considered are crucial almost everywhere. The book may be most useful, in particular, for those teaching mathematics.

The authors are by no means professional historians, so they apologize in advance for shortcomings from the point of view of the historian. To be specific, the authors tried to give a fair survey of the development of the subject without any claim of completeness. We hope that this book may encourage someone to take up a thorough investigation of the subject.

Although both subjects interrelate, one can read each contribution seperately.

We have only slightly touched on the latest developments; without considerably expanding the text, a full scale discussion was impossible. The interested reader is also referred to the literature.

We hope that our book will stimulate the interest of mathematicians of sundry backgrounds and that it may pass on some of the fascination that mathematics had for our predecessors.

Several people read our manuscripts and helped us with suggestions and comments for which we are grateful. In particular we thank H. J. M. Bos, U. Felgner and W. Marek.

Van Dalen
Monna

Set theory from Cantor to Cohen

Was beweisbar ist, soll in der
Wissenschaft nicht ohne Beweis
geglaubt werden.

R. DEDEKIND

Das Wesen der Mathematik liegt
in ihrer Freiheit.

G. CANTOR

GEORG CANTOR*

* The photograph of Cantor was kindley made available by the Universitäts Archiv of the Martin Luther Universität, Halle-Wittenberg, D.D.R.

Introduction

Set theory has dutifully performed the tasks that its founder Georg Cantor intended it to do and many more than Cantor could even dream of. In the present historical survey we have tried to trace some of the ideas and problems that were developed. Considering the scope of the lectures we had to restrict our attention to a modest part of set theory. We have choosen to follow set theory from Cantor via Zermelo, Fraenkel, Von Neumann, Gödel to Cohen and we hoped in this way to be faithful to the spirit of Cantor. As a consequence many subjects had to be excluded, among them other variants of axiomatic set theory (e.g. Quine's systems), the theory of types, topology, descriptive set theory (projective sets, etc.), hierarchy theory and many other subjects. Owing to the explosion in axiomatic set theory, following Cohen's fundamental papers, we could only superficially touch the recent results. A reader interested in these new methods and results should turn to literature.

For the reader interested in more information on set theory, there are many texts available, we will mention only some of these: Bernays-Fraenkel [8], Cantor [25], Cohen [30], Fraenkel – Bar-Hillel [39], Hao Wang [150], Van Heyenoort [61], Kneebone [70], Kuratowski-Mostowski [77], Mendelson [84], Mostowski [91], [92], Monk [87], Rubin [118], Quine [111] and the Proceedings of the AMS summer institute on Axiomatic Set Theory 1967 [109].

Forebodings

A history of set theory has certain advantages over comparable enterprises in other subjects. One of the reasons is that pure set theory is relatively young and it has sprung from a few, identifyable sources. Another reason is that the development of set theory mirrors the development of mathematical thought and practice in the twentieth century. Set theory was instrumental in the tremendous advance in abstract mathematics, nevertheless it brought mathematics almost to ruin with its embarrassing paradoxes. It also managed to polarise mathematicians into transfinitists (who gladly swallow the axiom of choice and more) and constructivists (some of whom deny sense to the second number class). One could say that no branch of mathematics did more to send the mathematician on an Odyssey, from which he has not yet returned[1]).

For the present-day mathematician, raised on New Math, it is hard to imagine how mathematics could manage so long without sets and set theory. The answer is that pre-Cantor mathematics did not need sets for its own purposes. Certainly, in geometry there were *loci*, in analysis sequences were studied, and probability theory had *events*, but one did not operate on these things (at least not beyond elementary operations given by 'and', 'or', 'not'), so there was no need for a general notion of set.

Looking back in history for phenomena that we associate with set theory, we notice that mathematicians and philosophers alike have been worried by infinity at a quite early stage. It would lead us too far to pay attention to that part of (pre-)history of set theory. We will have occasion to mention some older views in connection with Cantor. There is however one author, famous for quite different reasons, who dealt with the infinite in a very modern way. His name is GALILEO GALILEI. In his Discorsi [45] (1638) he reports the discussions of three gentlemen, Salviati, Sagredo and Simplicio. At the first day they come to discuss the continuous infinity etc. Salviati explains that there are just as many square numbers as numbers (positive integers are meant), because there are as many squares as there are square roots and each number is a square root (in our terminology the sets of natural numbers and squares of natural numbers are equipollent, or 'have the same cardinality'). The conclusion from this paradoxal phenomenon is given by Salviati: "The only solution I see is to say, infinite is the number of all numbers, infinite that of the squares, infinite that of the square roots; neither is the set of squares smaller than that of the numbers, neither is the latter larger; and finally, the attributes of equal, larger, smaller do not apply to the infinite quantities, they only apply to finite quantities". This strange phenomenon of a part not being smaller than the whole baffled everyone for a long time, so that the equipollence of the sets of natural numbers and the set of squares was referred to as the 'paradox of Galilei'. Maybe we should pay our respect to Galilei by coining it the theorem of Galilei[2]).

Much later a popular version of Galilei's theorem, which is ascribed to Hilbert, came

---

[1]) To continue the metaphor: on returning he might possibly find Penelope married to an intuitionist.

[2]) Actually the possibility of establishing a one-one correspondence between an infinite set and a proper subset has been noticed before among others by Plutarchus Proclus, Adam of Balsham (Parvipontanus), Robert Holkot.

into circulation. Suppose there is a hotel with a countable number of rooms (i.e. for every natural number n there is a room no. n) all of which are occupied. A new guest arrives and asks for a room. The owner of the hotel simply asks every guest to move to the next room, so the guest from no. 1 transfers to no. 2 etc. Now room no. 1 is vacant and the new guest can stay.

At another occasion countably many new guests arrive and ask for rooms while every room is occupied. This time the owner asks every guest, already present, to move from his room to the room with a number double the number of his present room. Now the guests move from room no. 1 to no. 2, from no. 2 to no. 4 etc. All the rooms with odd numbers are therefore vacant and every new guest can find a room.

Although most people were inclined to consider the infinite a suspect subject many mathematicians carried over procedures and facts from the domain of the finite to the domain of the infinite with a surprising mixture of courage and naivity. There is an instructive discussion between Leibniz and Johann Bernoulli ([79]) where Leibniz says:

> "Let us agree that there actually exist the segments of the line, denoted by $\frac{1}{2}, \frac{1}{4}, \frac{1}{8}, \ldots$, and that *all* members of this sequence exist, then you conclude that there is also an infinitely small member; in my opinion it only implies that actually every arbitrary, finite, definable fraction of every arbitrarily small size exists."

Bernoulli replied to this:

> "If ten members are available then necessarily the tenth member exists, if hundred, then the hundredth, . . ., therefore, if infinitely many members are given the infiniteth (infinitesimal) member exists."

Comparable attitudes, sometimes more, sometimes less crude, survived until long after Cauchy.

One of the first to analyse the notion of the infinite was BERNARD BOLZANO, a highly original and gifted mathematician, who unfortunately was somewhat isolated as a scholar. Bolzano wrote a monograph: "Paradoxien des Unendliches" [18] (published post-humously in 1851). As most expositions of a fundamental character in those days, it is partly polemical. The polemic passages, useful as they may have been, do not strike the present-day reader. The reader is likely to be far more impressed by those parts in which Bolzano expounds the fundamental notions. His terminology is somewhat cumbersome, but it is not hard to identify his terms in our terminology. Bolzano defines a collection (§ 3) (our translation for Inbegriff) as a whole consisting of certain parts (by 'parts' is meant 'elements'). He adds to this that some collections, although containing the same parts can be considered distinct, depending on the viewpoint from which they are viewed. As an example he mentions a drinking glass in one piece and one broken into several pieces. It is the way in which the parts form the whole, or the ordering that causes the distinction. One can discern here intensional aspects that were abolished only much later. Another interpretation of Bolzano's sophistication could be provided by assuming that Bolzano's collection is an ordered set, but this interpretation would presumably be too narrow.

If a collection is given by a property (Begriff), such that the ordering of the parts is irrelevant, then it is called a set (Menge). Thus we could say that a set is the extension of a property.

Bolzano defines a sequence of elements (we will now use 'elements' instead of 'parts') of a collection by stipulating that each element has a successor and that by a uniform law either the predecessor determines the successor or the successor the predecessor. Although this is an extremely deficient way of describing a sequence, we can surmise that Bolzano wanted to found the concept of sequence on the successor-relation only. Once the concept of sequence is available Bolzano introduces countable and finite sets (zählbare Vielheiten) and integers (ganze Zahlen) as sequences with a first element and with the property that every successor is obtained by adding a unit[3]) (of a given set) to its predecessor.

Note that this presupposes the correct notion of sequence, otherwise it would also cover ordinals. Also the notion of an infinite set can be introduced now: a set A is infinite if every finite set of elements of A is a proper subset of A (early authors did not use the adjective 'proper', but often 'subset' means 'proper subset'[4]). Or, as he elucidates, a set that cannot be exhausted by a finite sequence is infinite. Bolzano criticizes some definitions of infinite, e.g. something is infinite if it cannot be increased (Spinoza and others). Bolzano refutes this by the counterexample of a line bounded at one side and unbounded at the other. Such a half-line constitutes an infinite set, but it evidently can be increased. Also Hegel's concept of (mathematical) infinity, as being a variable magnitude does not escape Bolzano's attention. He objects to a philosopher claiming knowledge of an object, to which he assigns the predicate 'infinite' without establishing in any way a relation with infinite sets.

We will return to the 'variable magnitudes' when we discuss Cantor. Bolzano must be credited for recognizing the necessity to show the existence of an infinite set. He does not, as any newcomer would do, point to the integers or reals, but quite ingeniously mentions the set S of all sentences and truths. To show that S is infinite he defines the successor of a sentence A to be 'A is true'. Noticing that A and 'A is true' are distinct, he comfortably shows that S is infinite.

Not being content to give an example of an infinite set from the realm of the unreal

---

[3]) Units are objects (Gegenstände), they need not be urelements.
[4]) Dedekind however states that A is a subset of A [33].

he wants to give an example chosen from reality and this example he bases on the existence of God ([18] § 25).

1   The exploration of the new continent

In 1888 a remarkable treatise on the nature of our numbersystem appeared, its author was RICHARD DEDEKIND, a mathematician well-known for his contributions to number theory and algebra. It was certainly not the only monograph on the subject of Number, Kronecker and Helmholtz [75], [59] and others had already published treatises on the subject, but Dedekind definitely struck a new note. In his *"Was sind und sollen die Zahlen?"* Dedekind proposed to found the notion of number on the notion of set (or *system*, as he called it) in order to reduce arithmetic to something independent of intuitive observation (Anschauung), essentially to the direct consequences of the pure laws of thought. He opens the preface to the first edition with the memorable words "Was beweisbar ist, soll in der Wissenschaft nicht ohne Beweis geglaubt werden"[5]).

RICHARD DEDEKIND*

---

[5]) What is provable, must not be believed in science without proof.

* Reproduced from: Richard Dedekind — Gesammelte mathematische Werke. Herausgegeben von Robert Fricke, Emmy Noether und Ygstein Ore. Druck und Verlag von Friedr. Vieweg & Sohn Akt.-Ges. Braunschweig 1930.

His answer to the question posed in the title is: numbers are free creations of human mind. The great merit of Dedekind is that he does not stop at the stage of philosophical reflections on the nature of number, but that he carries out a by all means admirable program.

Dedekind begins his monograph with a systematic exploration of 'systems', which, in in the mind, are made up of objects (Dinge) with a property in common. The objects themselves are subjects of our thinking. Systems, as subjects of our thought, are again objects. It is this iteration of the process of collecting objects (or creating sets) that opens the way to an unrestricted set theory. Yet the abstract notion of a set is a rather bold one, as appears from Dedekind's remark that he will admit one-element systems (as if that were something special) and not the empty system, although he admits that under certain circumstances it may be advantageous to allow such a system. Clearly Dedekind strictly adheres to the idea that a system consists of elements, so if there are no elements, there is no system. Dedekind does not make explicit his objections to the empty system. Frege on the contrary allows concepts under which nothing falls and exploits these for the definition of the number 0 [4]. Frege, in his "*Grundgesetzen der Arithmetik*" [42], introduction, rightly critisizes Dedekind for not consistently identifying systems with concepts.

In a clear way Dedekind defines the well-known notions of union, intersection and subsystem[6].

A mapping (Abbildung) φ of a system S is defined as a *law* assigning objects to all elements of S, referring to Dirichlet's definition (cf. p. 89). Note that we are still far from the function as a set of pairs; the doctrine that 'everything is a set' required a considerable incubation period. Also one recognizes in this definition an intensional aspect, that later set theory eliminated.

One of the achievements, that still make Dedekind's name a household word in modern set theory, is the famous definition of finite and infinite. A system S is called infinite if it can be mapped by a one-one mapping onto a proper subsystem of itself[7], otherwise S is called finite. According to a footnote (§ 5) the idea was conceived before 1882 and communicated to Schwarz, Weber and Cantor. The beauty of the definition lies in its intrinsic character, it is formulated purely in terms of systems and mappings and there is no reference to natural numbers or other sophisticated objects. Of course with Dedekind a mapping is not a system so there is an extraneous element in the definition, which was however eliminated in due time. Dedekind, like Bolzano, proves the existence of an infinite set by exhibiting one: the totality of all (Dedekind's) thoughts.

Frege, who himself designed a foundational system for mathematics and logic, extensively criticized the psychological undertones in contemporary contributions to the

---

[6]   However with different terminology and notation.
[7]   The same definition was proposed by Pierce in 1885 [105].

foundations (cf. [42], preface), that so generously make use of 'human thinking', 'idea' (Vorstellung) etc.

Dedekind introduced a technical device, which he called *chain*, that enabled him to give smooth and short proofs. Let S be a set and φ a mapping of S into S. A subset $K \subseteq S$ is called a chain[8]) if $\varphi(K) \subseteq K$. The chain A_0 of a subset A is the intersection of all chains containing A. Dedekind recognizes in the notion of chain the two aspects: (i) the smallest set with . . . (by definition), and (ii) the inductive definition.

The latter definition gives the chain of A as the union of the successive images of A under φ. He connects (i) and (ii) by formulating and proving the appropriate *induction principle* ([33], 59). A set N is called simply infinite if there exists a $\varphi : N \to N$ such that N is the chain of one of its elements. Abstracting from the accidental character of N Dedekind obtains the natural numbers.

Although Dedekind left mathematics an extremely elegant foundation of arithmetic based on set theory, the development of pure set theory is the work of that highly gifted and original mathematician, Georg Cantor.

In the same year (1888) of the appearence of Dedekind's monograph, Giuseppe Peano published a book with the title "Calcolo geometrico secondo l'Ausdehnungslehre di H. Grasmann, predecuto dalle Operazioni della logica deduttiva ', in which he introduced the concepts of set theory that now belong to the standard part of any elementary text-book of mathematics. In particular the signs $\in$, $\cap$, $\cup$ are used in their present-day meaning.

The early history of CANTOR (cf. [25], [4], [85]) does not differ considerably from that of any contemporary mathematician[8a]). He studied mainly in Berlin under Kronecker, Kummer and Weierstrasz. At the age of 22 Cantor wrote his Ph. D. Thesis titled "De aequationibus secundi gradus indeterminati". Also his Habilitationsschrift dealt with arithmetic. Only after working on the uniqueness theorem for trigonometric series, Cantor came to deal with sets (of reals). Cantor, in a number of papers (1870-1872), extended a theorem on the uniqueness of the coefficients of trigonometric series, already considered by Riemann. In the course of his research Cantor conceived a new theory of irrationals (reals) and introduced the *derivative* of a point set. As this notion belongs to the domain of topology we will not dwell upon it. It will be sufficient to state that the derivative of a set is the set of its accumulationpoints. Cantor not only introduced the derivative, but boldly iterated the process of taking derivatives, thus entering the realm of the transfinite.

In this period Cantor met Dedekind on a trip to Switzerland. An intensive exchange of

---

[8]) Dedekind points out that the property of being a chain depends on φ, nowadays we should call K a φ-chain.

8a) Recently I. Grattan-Guiness published the following papers on Cantor: (i) The correspondence between Georg Cantor and Philip Jourdain, Jahresberichte DMV, 73, (1971) pp 111-130.
(ii) Towards a biography of Georg Cantor, Ann. of Science, 27 (1971), pp. 345-391

letters and some personal meetings followed. A number of letters are published in [100]. They make extremely interesting reading, as they give us a glimpse of the genesis of some of Cantor's ideas and Dedekind's insight. One can say that Cantor's new set theory started with his paper "*Über eine Eigenschaft des Inbegriffs aller reellen algebraischen Zahlen*"[9]) (1874).

In this paper Cantor shows that the set of real algebraic numbers is countable and that the set of reals is uncountable. Already in a letter of 29-11-1873 Cantor submits to Dedekind the conjecture that the set of reals is uncountable. Dedekind answers that he can neither prove nor disprove the conjecture, but he adds that the problem does not deserve so much attention, as it has no practical applications anyway. Later Dedekind writes in his notebook that Cantor's proof of the existence of transcendent real numbers on the basis of the conjecture (by then a theorem) has convincingly refuted the last part of his answer.

Actually Cantor proved the following theorem: *let a countable set S of reals be given, then we can find in every interval (a,b) a real number not belonging to S.*

The proof is sufficiently interesting to be reproduced here.

Consider the set S with elements s_1, s_2, s_3, ... (i.e. we have an enumeration of S). Let a and b be given, such that $a < b$, find the first two s_i and s_j such that $a < s_i < b$ and $a < s_j < b$. Let $s_i < s_j$ and put $a_1 = s_i$, $b_1 = s_j$. Now suppose that a_1, ..., a_n, b_1, ..., b_n have already been defined. Then we locate the first s_i and s_j such that $a_n < s_i$, $s_j < b_n$, if there are such elements in S. If not, the proof is finished because the interval (a_n, b_n) contains at most one point of S.

So let us suppose that there is an infinite sequence of intervals (a_n, b_n). There are two possibilities:

(i) $\lim a_n = \lim b_n = r$, then $r \notin S$ because otherwise $r = a_k$ or $r = b_k$ for some k, which is impossible as $a_k < r < b_k$ for all k.

(ii) $a = \lim a_n < \lim b_n = b$. In that case any point between a and b satisfies the requirement.

The above proof depends on the topology of the real line. It is no exaggeration to hail the 1874 paper as a milestone in the history of mathematics. Here for the first time the uncountability of a set was established. Earlier, only the vague distinction finite – infinite was known, but here we already have two different kinds of infinity!

Probably, countability problems have been considered before, although they must have been considered as amusing problems for recreation.

In [123] it is reported that Weierstrasz at a seminar in 1873/74 asked for an enumeration of the rationals, remarking that once a student had invented one. Schoenfliess remarks that, on account of the evident negative attitude of the seminar, nothing resulted.

---

[9]) On a property of the totality of all real algebraic numbers.

Meanwhile Weierstrasz was interested in Cantor's ideas and Cantor explicitly refers to him in his paper [26].

Cantor's next spectacular enterprise was the establishment of a one-one correspondence between the n-dimensional euclidean space R^n and the 1-dimensional space R (i.e. the real line). It is interesting to read the history of this theorem in the Cantor-Dedekind letters. Cantor communicates a proof of this theorem to Dedekind in the letter of 20-6-1877, two days later Dedekind answers Cantor and points out a mistake in the proof, due to Cantor's use of decimal expansion of reals (the proof was repaired by König at a later occasion).

Cantor acknowledges the mistake and sends a correct proof to Dedekind the 25th of June, this time he uses continued fractions instead of decimal expansions.

Cantor had apparently discussed the problem with a friend in Berlin ([100], letter of 18-6-1874), who declared the matter to be absurd, as evidently two independent variables could not be reduced to one variable!

Cantor rejoiced so immensely in the result, that he wrote to Dedekind (25-6-1877) "It appears to me that all philosophical or mathematical deductions, based on the fallacy of the invariance of dimension, are inadmissable". Dedekind, being a solid mathematician, not subjected to rash conclusions, warns Cantor (2-7-1877) not to indulge in polemics concerning the invariance of dimension because the intended correspondences between continuous manifolds (stetige Mannigfaltigkeiten) should be continuous. Cantor agrees with his friend and sets out to prove the invariance of dimension; he presents a proof in the paper "Über einen Satz aus der Theorie der stetigen Mannigfaltigkeiten" (1879). Unfortunately the proof was fallacious (a correct proof was provided by L. E. J. Brouwer in 1910 [22]), the problem turned out to be so hard that its solution had to wait until another genius came along.

Another problem that Cantor posed in this period underwent the same fate: the *Continuum Hypothesis* was formulated in "Über unendliche lineare Punktmannigfaltig-keiten", Nr. 5, § 10 ([25], p. 192, p. 244), where Cantor says that he hopes to provide soon an exact proof of the equivalence (gleichmächtigkeit) of the interval $(0,1)$ and the second number class (in our terminology $2^{\aleph_0} = \aleph_1$). Also Cantor formulates a 'higher' continuum hypothesis, namely that the set of all real functions has the cardinality of the third number class. The class of continuous (resp. integrable) functions however is claimed to be of the cardinality of the second number class (no proof) ([25], p. 297).

In the same series of papers, which contain a wealth of new ideas and concepts, we also meet the *Cantor-Bernstein theorem* ([25], p. 201). Cantor states the theorem for sets of cardinality of the second number class and concludes it as a corollary of the following theorem: a subset of the second number class is (i) finite, or (ii) countable, or (iii) equivalent to the second number class. The proof of the latter theorem essentially uses the wellordering of the second number class. The general Cantor-Bernstein theorem is asserted by Cantor without proof. Cantor's formulation of the theorem reads: *let*

11

$M'' \subseteq M' \subseteq M$ and let there be a one-one mapping of M onto M'', then there is a one-one mapping of M onto M'.

Also in his voluminous paper "Beiträge zur Begründung der transfiniten Mengenlehre" (1895-97) Cantor does not prove the Cantor-Bernstein theorem, but instead remarks that it is a consequence of the comparability property of cardinals. He explicitly points out that the latter problem is open. Compare also the letter to Dedekind of 5-11-1882 [100].

The comparability of cardinals turned out to be equivalent to the axiom of choice [55].

The Cantor-Bernstein theorem has a remarkable history. Dedekind already in 1887, gave an elegant short proof of the theorem (see [34], p. 447-449). This happened before Cantor published his version. Apparently Dedekind forgot all about this note and the theorem. J. Cavaillès found the note among the posthumous papers. Curiously enough Dedekind gave another proof in a letter to Cantor (29-8-1899), apparently unaware of the fact that he had seen the problem before ([25], p. 449). This letter refers to a visit of F. Bernstein (der junge Herr Felix Bernstein), who mentioned the Cantor-Bernstein theorem to Dedekind, upon which the latter conjectured (to Bernstein's astonishment) that its proof schould be easy be means of the methods of "Was sind und sollen die Zahlen?".

Much later Zermelo published a proof of the Cantor-Bernstein theorem in the Mathematische Annalen (1908), that was essentially Dedekind's proof. Neither Cantor nor Dedekind seem to have advertised the proof. In the meantime in 1896 Schröder had published a fallacious proof, independently Bernstein had presented a correct proof in a seminar at Halle (where Cantor was teaching). Bernstein's proof appeared in Borel's "Leçons sur la theorie des fonctions" in 1898 [19] (p. 103 ff.).

Finally it must be remarked that Mittag-Leffler in a letter of 5-4-1883 formulates the Cantor-Bernstein theorem as a conjecture. The Cantor-Bernstein theorem has never stopped to attract the attention of mathematicians, e.g. constructive versions have been considered, the best-known of which is Myhill's theorem in recursion theory [115], p. 85 (see also [148], p. 104, [31]).

Cantor's innovations must have evoked considerable resistence, not merely with mathematicians (like Kronecker), but also with philosophers, theologists and other scholars, who traditionally dealt with the infinite. Cantor, who was well-read in philosophy and theology, defended his views vigorously in a number of papers, e.g. "Grundlagen einer allgemeine Mannigfaltigkeits lehre"[10]) (1883), "Über die verschiedene Standpunkte in bezug auf das aktuell Unendliche"[11]) (1885, 1890), "Mitteilungen zur Lehre vom Transfiniten"[12]) (1887, 1888).

---

[10]) Foundations of a general theory of manifolds.
[11]) On various viewpoints concerning the actual infinite.
[12]) Communications on the theory of the Transfinite.

In the mathematical world, opposition against Cantor was lead by Leopold Kronecker, a brilliant number theoreticist and algebraicist, who advocated a purge of mathematics. Considering his opinion, that only integers (and objects constructed from these, e.g. rationals) are eligible for meaningful mathematics (his slogan "God made the integers, all the rest is the work of man" is well advertised), it is not surprising that he deeply depreciated Cantor's excursions into the Transfinite. Kronecker had a well established position in Berlin and there are indications that he actually campaigned against Cantor and his set theory. Cantor, who was never promoted from the university of Halle to one of the prestigeous universities (e.g. Berlin), was extremely sensitive to the lack of appreciation of his work in Germany and the (real or supposed) machinations of mathematicians like Kronecker bitterly disappointed him. His relation to Kronecker provides an example of an academic dispute that ruined personal relations. Despite attempts of reconciliation by Cantor (cf. [122]) normal relations were never again established.

Where Kronecker's guerilla was an extreme case of opposition to Cantor's 'new math', many mathematicians considered set theory with its preference for the transfinite a speculative affair. No less an authority than Gauss had given vent to the 'horror infiniti' and thus set the tone. In a letter from Gauss to Schumacher ([45], p. 216 f.) we read:

> "... so protestiere ich zuvörderst gegen den Gebrauch einer unendliche Grösse als einer Vollendeten, welcher in der Mathematik niemals erlaubt ist. Das Unendliche ist nur eine 'façon de parler', indem man eigentlich von Grenzen spricht, denen gewisse Verhältnisse so nahe kommen als man will, während andern ohne Einschränkung zu wachsen verstattet ist"[13]).

Gauss bases his judgement on views concerning man and his knowledge, as appears from the same letter:

> "... der endliche Mensch sich nicht vermisst, etwas Unendliches als etwas Gegebenes und vor ihm mit seiner gewohnten Anschauung zu Umspannendes betrachten zu wollen."[14])

If mathematicians hesitated to accept the actual infinity, philosophers, mostly vaguely acquainted with contemporary mathematics, were defenitely shocked. A considerable part of Cantor's papers "Über unendliche lineare Punktmannigfaltigkeiten" and "Mitteilungen zur Lehre vom Transfiniten" is used to counter objections of philosophers.

---

[13] I protest first of all against the use of an infinite magnitude as something finished, which never is allowed in mathematics. The infinite is but a manner of speaking, as one properly speaks of bounds, that certain proportions approximate arbitrarily close, while some are allowed to increase unlimitedly.

[14] ...finite man should not assume to consider something infinite as something given and comprehensible by means of his customary intuition.

Most of these objections concerned the notion of infinity; the majority of Cantor's contemporaries adhered some version of the potential infinite.

Unfortunately, lack of mathematical background and precision made them an easy prey for the well-read Cantor. It would lead us too far to examine the controversies thouroughly. We will just select one specific case, namely the review of Cantor's "Grundlagen einer allgemeinen Mannigfaltigkeitslehre" by L. Ballauf, a philosopher from Herbart's school [3].

Ballauf's review does not at first sight differ from others, but after Ballauf mentions both the notion of potential and actual infinite and remarks that if one conceives e.g. the set of natural numbers as given and completed, nothing is wrong with the actual infinite, an unusual thing happens: the editors step in to suppress this heresy. Cantor's comments on this occasion are instructive and amusing ([25], p. 392 ff.): "In my opinion the reviewer, who on a number of occasions has shown to comprehend my thoughts well, has been forced by terrorism of the leaders of the school (i.e Herbart's) to take a much stronger stand, than conforms to his own views. This strikingly appears, where the editors suddenly cut into his open-minded, free reflections, to carry the poor man back to the dark subterranean prison of Herbartian Dogmatics, where no redeeming ray of light can penetrate". Cantor's defense against the Herbartians (and others) is summed up in two points: (i) they define infinite as potential infinite and hence conclude that no actual infinite exists; (ii) the definition of the infinite (apparently an infinite magnitude is intended) presupposes an actual infinite domain. The concept of the infinite, according to Herbart's school, is based on a variable bound that can (resp. must) be moved forward. This variable bound, says Cantor, can only be moved thanks to the actual infinite road. The reader should consult Cantor's paper to perceive the care with which Cantor tries to answer the many objections, from various quarters, to his theory of infinite sets and magnitudes.

In mathematical circles however these activities of Cantor were considered suspect or worse. The following extract from the letter of 17–10–1887 from H. A. SCHWARZ to C. WEIERSTRASZ (in [85] p. 255) may illustrate this:

> "After I found an opportunity to contemplate this note[15]) I cannot conceal that it seems me to be a sickly confusion. What in the world have the churchfathers to do with the irrational numbers?! . . . If only one could succeed to accupy the unhappy young man with concrete problems, . . ., otherwise he will certainly come to a bad end."

The set theoretical apparatus used by the present-day working mathematician is almost exclusively the heritage of Cantor. The greater part of the set theoretical terminology

---

[15]) Mitteilungen zur Lehre vom Transfiniten [25], p. 378 ff.

goes back to Cantor, who, in a letter to L. Scheeffer [122] declared that with respect to the introduction of new notions: "I am extremely careful with the choice of those, as I take the position that the development and propagation in no small degree depends on a fortunate and properly fitting terminology".

Not considering the many topological notions that Cantor introduced, we may list a few important ones: well-ordering, ordinal, cardinal, diagonalprocedure.

The notions of well-ordering and ordinal (Cantor initially uses 'number' (Zahl) and later 'ordernumber' (Ordnungszahl), are introduced in his paper "Grundlagen einer allgemeinen Mannigfaltigkeitslehre" (1883). Cantor generates the ordinals by two generation-principles, the first principle adds one element to a previously obtained sequence of ordinals, and the second principle adds an element as supremum of an infinite sequence of previously obtained ordinals. So he obtains

$1, 2, 3, \ldots, n$
$1, 2, 3, \ldots, n, \ldots, \omega$
$1, 2, 3, \ldots, n, \ldots, \omega, \omega+1, \omega+2, \ldots, \omega+n$
$1, 2, 3, \ldots, n, \ldots, \omega, \omega+1, \omega+2, \ldots, \omega+n, \ldots, 2\omega$
etc.[16]).

The *first number class* is the set of all ordinals of the form n (i.e. all ordinals less than ω) the *second number class* is the set of all countable ordinals. It is shown that the second number class itself is not countable. Also addition and multiplication of ordinals are introduced.

In the papers of 1895-1897 (the *Beiträge* for short) the body of set theory is systematized. The very first sentence of the Beiträge is Cantor's definition of a set: "By a 'set' we understand every collection M of certain well-defined objects m of our observation or of our thinking into a whole".

There is a report of Bernstein ([34], p. 449) on Dedekind's and Cantor's intuition of the notion of a set, that may be illuminating.

Dedekind thought of a set as a closed bag, containing certain well-determined objects, that one could not see and of which nothing was known but their existence and determinedness. Cantor on the other hand on some accasion stated that he thought of a set as an abyss.

The above concise definition certainly is closer to the bag than to the abyss, but the occurrence of the paradoxes lent set theory an abysmal aspect; we will return to it later.

The notion of *cardinal number* is introduced in the not unusual form of abstraction:

"The cardinal number of M is the general concept (Allgemeinbegriff), which by means

---

[16]) 2ω is Cantor's original notation. Nowadays we write $\omega 2$ instead of 2ω.

of our active thinking-faculty results from the set M by abstracting from the nature of the elements and the order in which they are given".

To express this double abstraction the cardinal number of M is denoted by $\overline{M}$.

By our standards this definition of set is not a mathematical one, indeed it is hard to prove facts on the basis of this definition. For example the proof that equivalent sets have the same cardinality[17]) and vice versa ([25], pp. 283, 284) rests on 'insight'. Even Cantor takes liberties with his notions, he boldly asserts the equivalence of M and $\overline{M}$. But since equivalence is defined by the notion of one-one correspondence, $\overline{M}$ has to be a set, which was not intended by 'general concept'. The elaborate definition of cardinal shows that Cantor did not have the machinery of the equivalence classes at his disposal, it also shows that Cantor had not yet reached the level of abstraction that set theory attained later.

The reference to 'the order in which the elements are given' indicates that Cantor had at the back of his mind a somewhat constructive notion of set.

The same remarks apply to the definition of *order type*.

Well-ordered sets are defined as ordered sets with the following properties: (i) there is a first element and every element, except for the last one, has an immediate successor, (ii) if a subset has an upper bound in the set, then it has a least upper bound. This definition does not essentially differ from the definition given in 1883 ([25], p. 168).

The property by which we characterize well-ordered sets, i.e. every non-empty subset has a first element, is derived by Cantor. Cantor's formulation differs from the usual one insofar that Cantor deletes the predicate 'non-empty'. As a matter of fact the empty set does not occur in Cantor's work, e.g. if a pointset contains no accumulation points, it has no derivative according to Cantor ([25], p. 140). Where Dedekind explicitly excludes the empty sets, Cantor does not even mention it, I have not been able to find out why[18]).

Although in modern set theory the well-ordering of a set is defined by the first element property, the earlier definition of Cantor has survived in constructive mathematics. In particular Brouwer has isolated the constructive content of Cantor's definition and built an intuitionistic theory of well-ordering on it [24]. The Beiträge is a paper that even today is well worth reading, it treats the main topics of the theory of cardinals, order-types, ordinals in a masterly way.

The theory of ordinals is restricted to the second number class, there is no motivation for this restriction. It is clear that Cantor had considered a wider class; according to Bernstein [10] Cantor already in 1895 knew the paradox of all ordinals (thus before

---

[17]) We will use "cardinal number", "cardinal", "cardinality", indiscriminately.

[18]) Even in the second edition of Schoenflies textbook [121] the empty set does not occur (cf. p. 34 footnote 2). The 0 in $P^{(\nu + 1)} = 0$ at p. 258, is just a symbol, meaningful only in this context.

Burali-Forti). This fact may explain Cantor's caution, but there are no indications to confirm it.

Among the multitude of new results and procedures introduced by Cantor we will single out a few.

The *diagonalprocedure*. The German association of mathematicians (Deutsche Mathematiker Verein) was founded in 1890 and one of its active advocates was Georg Cantor, who became the first chairman. Notwithstanding the strained personal relations, Cantor invited Kronecker to address the first meeting in 1891 at Halle. When Kronecker, after the death of his wife, excused himself[19]), Cantor gave a talk instead. In this talk a simple proof of the uncountability of the reals was presented[20]).

The method was the now famous diagonalprocedure, given both for countable sequences of two elements m and w and for characteristic functions in general.
Let us sketch here Cantor's famous argument:

Let L be a set and M the set of all functions f on L with values 0 and 1 only $(f : M \to \{0,1\})$. Clearly $L \underset{1}{\leqslant} M^{21}$), because of the mapping φ with $\varphi(x) = c_x$ (where c_x is the constant mapping with value x).

Now suppose $L \underset{1}{=} M$ by the mapping ψ. Then the mapping $\Psi : L \times L \to \{0,1\}$ is defined by $\Psi(x,y) = (\psi(x))(y)$ (i.e. first apply ψ to x, the result $\psi(x)$ is a mapping of L into $\{0,1\}$, then apply $\psi(x)$ to y, the result is 0 or 1).

The mapping Ψ_d, defined by $\Psi_d(x) = \Psi(x,x)$, maps L into $\{0,1\}$ (the diagonal mapping). Now consider the contra-diagonal mapping Ψ_{cd}, defined by $\Psi_{cd}(x) = \begin{cases} 0 \text{ if } \Psi_d(x)=1 \\ 1 \text{ if } \Psi_d(x)=0 \end{cases}$

$\Psi_{cd} \in M$, hence $\Psi_{cd} = \psi(x_0)$ for some x_0.

Now consider the value of Ψ_{cd} in x_0:

$$\Psi_{cd}(x_0) \neq \Psi_d(x_0) = \Psi(x_0,x_0) = (\psi(x_0))(x_0) = \Psi_{cd}(x_0).$$

Here we have reached a contradiction, therefore $L \underset{1}{<} M$ Q.E.D.

The proof is Cantor's, with some changes in notations. Indeed Cantor makes essential use of set theoretical notions (e.g. mappings operate on mappings etc.) beyond the mere manipulation of Boolean operations.

---

[19]) See the letter to Cantor [76].

[20]) Published in "Über eine elementare Frage der Mannigfaltigkeitslehre" [25] p. 278 ff.

[21]) $L \underset{1}{\leqslant} M$ stands for "there is a one-one mapping of L into M"

$L \underset{1}{<} M$ stands for "$L \underset{1}{\leqslant} M$ and not $M \underset{1}{\leqslant} L$"

$L \underset{1}{=} M$ stands for "there is a one-one mapping of L onto M"

An alternative formulation of the above result is: the cardinality of the set of subsets of L (the powerset of L) in greater than L.

A corollary, noted by Cantor, is "there is no maximal cardinality". The diagonal procedure, novel as it appears in Cantor's proof, had already been applied by P. DU BOIS-REYMOND in 1875 [17] in his proof that there is no function with a minimal eventual increase to infinity. To be precise:
there is no sequence of real functions $\lambda_1, \lambda_2, \lambda_3, \ldots$ with the property that

$$\lim_{x \to \infty} \frac{\lambda_r(x)}{\lambda_{r+1}(x)} = \infty,$$

and such that for each φ with

$$\lim_{x \to \infty} \varphi(x) = \infty,$$

we have

$$\lim_{x \to \infty} \frac{\varphi(x)}{\lambda_k(x)} < \infty$$

for some k. Du Bois-Reymond applied diagonalisation by considering the functions λ_k in the intervals $[k-1, k]$ and pasting them together (after shifting them if necessary). Only after Cantor's application the diagonal procedure became a household item for mathematicians. Especially in logic and recursion theory the diagonal procedure became one of the main tools.

WITTGENSTEIN, [151] Teil I, Anhang II, takes an altogether negative view of the diagonal procedure, because it promises more than it can deliver. According to him the problem whether the real numbers can be ordered into an ω-sequence does not make sense because there is no precise idea of the ordering of all real numbers. All that is left of the diagonal procedure after Wittgenstein's criticism is a method for the calculation of a real number distinct from a given sequence of real numbers. From this position to Cantor's is but one step, a step which, however, Wittgenstein refutes as a meaningless play of words.

The back-and-forth method. In the Beiträge [25] p. 304, a remarkable property of ordered countable sets was discovered, which lent an exceptional role to the set of rational numbers. In the following discussion of that property we will use modern terminology.

Two ordered[22] sets M and N are called isomorphic if there is a one-one mapping f

---

[22] "Ordered" means "linearly ordered".

from M into N that preserves the order, i.e. $x \underset{M}{<} y \Leftrightarrow f(x) \underset{N}{<} f(y)$; the map f is called an isomorphism.

The *order-type* of an ordered set is the equivalence class of all sets isomorphic to it.

Now there are many non-isomorphic ordered sets (not surprisingly). It would be interesting to characterize by a description in terms of order and cardinality a class of ordered sets, such that they all have the same order-type. Here is an example that fits the requirements: "there are exactly two elements".

Evidently every two ordered sets with two elements are isomorphic. Now an example that does not work: "there is a first and a last element". The set $\{0,1\}$ and the (closed) interval $[0,1]$ satisfy the sentence, but they are not isomorphic.

Cantor found a non-trivial example of a description that determines countable ordered sets up to isomorphisms: "There is no first element, no last element. Between every two elements lies at least one element".

Ordered sets satisfying these requirements are called *densely ordered sets without endpoints*.

By an ingenious method Cantor established an isomorphism between any two countable densely ordered sets without endpoints, the procedure is illustrated below, using the figure.

The sets A and B are enumerated (without repetitions) in sequences $a_0, a_1, a_2, \ldots$; $b_0, b_1, b_2, \ldots$

The isomorphism is constructed stepwise

step 1: associate a_0 to b_0

step 2: locate a_1 with respect to a_0. As $a_1 > a_0$, associate a_1 to an element greater than b_0, e.g. b_2

step 3: locate b_1 with respect to b_0 and b_2, if it is not associated to any a, choose an

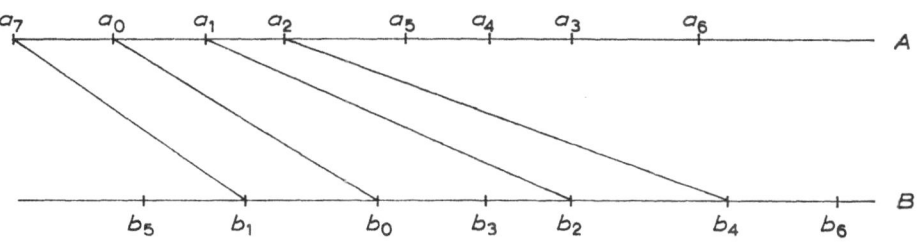

Figure 1.

19

element of A with the same location with respect to a_0 and a_1. In our case $b_1 < b_0$ and $b_0 < b_2$, so we associate a_7 to b_1 (because $a_7 < a_0$ and $a_7 < a_1$). If b_1 were already associated to an a_i, we would have gone on to step 4

step 4: check whether a_2 has already been associated to some b_j, if not locate a_2 with respect to all 'earlier' a_i's and choose a b_k that has the same location with respect to the corresponding 'earlier' b_j's, e.g. take b_4

step 5: check whether b_2 has been associated. The answer is yes, so we go on to

step 6: check whether a_3 has been associated, if not locate it with respect to all 'earlier' a_i's and choose a b_k with the same location with respect to all 'earlier' b_j's, e.g. take b_6.

etc.

It is clear that this process exhausts A and B and that the associated pairs determine an isomorphism. The main trick in the construction of the isomorphism consists of the going back and forth between A and B.

Cantor's result has been translated in modern terminology as follows: the theory of densely ordered sets without endpoints is $\aleph_0$-categorical. A specific set of the mentioned class is the set of rationals. Cantor denoted its order-type by η. The back-and-forth method has been applied with succes in model theory (cf. [5]).

Normal form for the second number class ([25] §19). An important result, giving insight into the second number class, was the reduction of every α to the form $\omega^{\alpha_0} a_0 + \omega^{\alpha_1} a_1 + \ldots + \omega^{\alpha_k} a_k$ where $\alpha \geqslant \alpha_0 > \alpha_1 > \ldots > \alpha_k$, $1 \leqslant a_i \leqslant \omega$, $0 \leqslant k \leqslant \omega$ where the addition, multiplication and exponentiation are defined by induction[23]).

The ε-numbers. The roots of the equation $\omega^\xi = \xi$ are called the ε-numbers. The first of these ε-numbers is $\varepsilon_0 = \omega^{\omega^{\omega^{\cdots}}}$, an ordinal that played an essential role in the consistency proofs for arithmetic. ε_0 is the supremum of the ordinals for which one can prove the principle of transfinite induction in arithmetic (cf. [125], [46]).

The Cantor space. (or Cantor set). In a footnote to the paper "Grundlagen einer allgemeinen Mannigfaltigkeitslehre" Cantor gave his famous example of a perfect set which is nowhere dense, consisting of all real numbers of the form $\sum_{i=1}^{\infty} \frac{c_i}{3^i}$ (where $c_i \in \{0,2\}$). This set plays an important role in foundational work. For example it can be considered as the set of all characteristic functions of the set of natural numbers,

---

[23]) Cantor defines the first two operations by means of sums, the last one by induction [25], p. 321, 322, 336.

hence as the set of all subsets of the set of natural numbers. Also in topology the Cantor space is a convenient tool.

The alephs. Cantor introduced a special notation for the cardinal-numbers of well-ordered sets. In the Beiträge only the first two alephs $\aleph_0$ and $\aleph_1$ (being the cardinals of all finite numbers and the cardinal of the second number class) are elaborated. The higher alephs are mentioned there and also in the letter from Cantor to Dedekind of 28-7-1899, they essentially are the cardinals of initial ordinals[24]) (i.e. ordinals that are not equivalent to smaller ordinals, cf. [77], p. 279).

The period we considered in this section can be characterized as the years of exploration. Many new phenomena were discovered by the courageous and imaginative Cantor and the solid Dedekind. Doubtless the father of set theory is Georg Cantor, who attacked the problems of infinity singlehanded.

The attitude of the mathematical establishment with respect to this new discipline is hard to assess. We already know the hostile reaction of Kronecker, but there also was the sympathetic appreciation of Mittag-Leffler[25]).

By keeping track of the status of set theory in the Jahrbuch der Fortschritte der Mathematik [26]) we more or less will arrive at a picture of the evaluation of set theory in mathematics.

In 1894 Cantor is reviewed in the section "Principles of geometry", the next year we find set theory under the head of "Philosophy and education". In 1905 set theory is given a separate section, but still under philosophy. Finally in 1919/20 set theory gains independence and is no longer classified as a subdiscipline. As a matter of fact the boundaries between the different aspects of set theory were rather vague, topology for example was not easily distinguishable from pure set theory. It is worth mentioning that Cantor's work in the Fortschritte was mostly reviewed by Vivanti, a sympathetic mathematician who himself was active in set theory. It certainly was a stroke of good luck that Cantor's articles were not crushed by somebody like Kronecker.

2   The paradoxes

As Bernstein reports [10], Cantor was already in 1895 aware of the paradox of the set of

---

[24]) Kowalewski ([73], p. 202) mentions Cantor's use of initial ordinals in his lectures.

[25]) There is however a curious incident in the otherwise friendly relation between Cantor and Mittag-Leffler. For this we refer to the delayed publication in 1970 of a paper written by Cantor in 1884, commented upon and edited by I. Grattan-Guiness [51].

[26]) The German mathematical reviews of those days.

all ordinals; also the paradox of the set of all sets must have been known to him at that time (cf. [34], p. 449). Cantor wrote to Dedekind in the summer of 1899 on the subject of the contradictions that seemed inherent to some sets [25], p. 443 ff.

Cantor discusses the 'system' Ω of all ordinals. If Ω were a set, it would have an ordinal δ, which would be greater than all ordinals α, in particular δ would be greater than δ, which is absurd. These and like considerations led Cantor to split 'multitudes' (Vielheiten) into two categories: (i) those that can be collected into a unit (a completed thing), (ii) those that cannot be conceived as a completed object without giving rise to contradictions. He called these *consistent* and *inconsistent multitudes*. The crucial point is that some classes of objects cannot be collected (or thought together) into a new object. We will meet the same distinction with Von Neumann, who distinguishes *sets* and *classes*.

Cantor indicates more inconsistent multitudes, e.g. the collection of alephs[27]), the collection of all sets. Cantor's proof of the inconsistency of the 'set of all sets' makes use of properties of cardinals and unions ([26], p. 448). We will sketch a simplified proof here.

Let S be the set of all sets and let P be the set of all subsets of S, then according to Cantor's theorem $S \underset{1}{<} P$ (cf. p. 17). But every element of P is a subset of S, i.e. a set, hence an element of S. Therefore $P \subseteq S$ and so $P \underset{1}{\leqslant} S$ holds. This contradicts Cantor's theorem.   Q.E.D.

Cantor inadvertently makes use in these letters of principles equivalent to the axiom of choice, but he asks Dedekind to try his hand at the comparability of sets for which Schröder, Bernstein nor he himself have been able to provide a proof. He refers to an indirect proof (which was fallacious) in an earlier letter.

All this went on in private correspondence and discussions, how much leaked to the mathematical world is hard to ascertain. Anyway in 1897 CESARE BURALI-FORTI published his paper "Una questione sui numeri transfiniti"[28]) in the Rendiconti del Circolo matematico di Palermo. What we now call the Burali-Forti paradox is the property that the set of all ordinals is well-ordered, hence has (is) itself an ordinal, which must be at the same time a member of the set and greater than all elements of the set.

Burali-Forti used the obtained contradiction to show that the set of ordinals is partially ordered and not (linearly) ordered. Cantor's proof that the ordinals are linearly ordered still had to appear then in the Beiträge. The paper of Burali-Forti is based on a notion of well-ordering that is actually weaker than Cantor's notion. In a supplementary note he later corrected his notion. The derivation of the contradiction however carries over to

---

[27]) The alephs $\aleph_0$, $\aleph_1$, ..., $\aleph_a$, ... can be considered the cardinalities of the ordinals. We will return to them later.

[28]) A problem on transfinite numbers. English translation in van Heyenoort [61].

Cantor's ordinals. Cantor was not pleased by Burali-Forti's paper; in a letter to Young [152] he writes: "What Burali-Forti produced is thouroughly foolish. If you go back to his paper in the Circolo Mathematico, you will remark that he has not even understood properly the concept of a well-ordered set".

Burali-Forti derived his 'paradox' as follows: if the set Ω of all ordinals is well-ordered, then it is itself an ordinal. Clearly for each ordinal α we have $\alpha < \Omega$, in particular $\Omega + 1 < \Omega$, which is absurd. Burali-Forti correctly uses the fact that each ordinal, including Ω, has a successor. It is instructive to read Young's reaction on Burali-Forti's paper. Commenting on the use, made of Ω, Young writes "Now, says Burali-Forti, take any set which is well-ordered and has type Ω, and place after it a new element. Etc.". When Burali-Forti reduces the existence of the set Ω ad absurdum, Young exclaims:

"Of course it is absurd. It is as absurd as to say 'Take a cat that is a dog' or 'take an even prime number greater than 3'. When you have taken in your mind a set of type Ω, you have taken everything, nothing remains to give a new element (it might be contended that we do not need to take a new element; it would suffice to take the first of our numbers and place it after the last, so *after* Ω. But the fallacy is really the same as above, *there is no place after* $\Omega, \ldots$) and the whole of the reasoning is mere confusion".

I quoted Young at length in order to demonstrate that even mathematicians of good standing, like Young, who himself worked in set theory, were at a loss when faced with the witchcraft of inconsistent sets. In Young's case there also was a certain amount of partizanship which made him conclude "... and no writer on the Theory of Functions of a Real Variable ought to leave his readers to suppose that Burali-Forti or anyone else, in Peanese[29]) or any other strange tongue, has proved the foundations of Cantor's theory to be unsound".

Burali-Forti wrote his paper using the symbolism introduced by Peano, therefore it is not easy to read. The result, however, echoed through the mathematical world. We find for example a discussion of the paradox in the book "Science et Méthode" of Henri Poincaré, one of the foremost mathematicians of the day, who had a keen interest in the foundations of mathematics. Poincaré was extremely sceptic with respect to the new discipline of symbolic logic and he did not always distinguish clearly between symbolic logic and set theory. Burali-Forti's choice of notation may have led him to suspect connections between the two. The appearance of the paradoxes made him rejoice. In the section Mathematics and Logic [107] he discusses the Burali-Forti paradox with Hadamard. Hadamard (correctly as we agree now) concludes that it is illegal to collect all ordinals into a set. Poincaré encounters this solution by pointing out that Ω is definable in Peano's

---

[29]) Poincaré's nickname for Peano's symbolic language.

system, he continues: "I would like to know who could prevent him (to define Ω) and can one say that an object does not exist if one calls it Ω?" Peano's formalization of mathematics (including set theory) clearly was premature, the actual subject of set theory was not far enough explored to harness it already in the frame of a formal theory.

Hadamard, from the viewpoint of naive set theory, certainly was justified in his rejection of the set of all ordinals, but Poincaré was equally justified in demanding the logicians to stick to their own rules. One cannot but conclude that Poincaré again scored a point against symbolic logic, and set theory unfortunately shared in the blame.

One might think that intricate notions like cardinal, well-ordering or ordinal caused the contradiction, this believe, however, was dispelled by BERTRAND RUSSELL, who around the turn of the century got deeply interested in the foundations of mathematics. After the International Congress of Philosophy in 1900 Russell started reading and studying the foundational papers of Peano. The latter had impressed Russell by a precision, not possessed by others.

In his autobiography (volume I, p. 144) Russell writes that

> "The Congress was a turning point in my intellectual life, because I met there Peano. I already knew him by name and had seen some of his work, but had not taken the trouble to master his notation. In discussions at the Congress I observed that he was always more precise than anyone else, and that he invariably got the better of any argument upon which he embarked. As the days went by, I decided that this must be owing to his mathematical logic."

In cooperation with Whitehead, Russell advanced rapidly in the new logic. However this period of honeymoon delight, as Russell expressed it ([120] p. 13), ended in June 1901. At that time the rumours of Cantor's paradoxes, in particular the one of the greatest cardinal, reached Russell, it made him consider the set of all things and the sets which are not a member of itself. The latter one gave rise to the *Russell paradox*. Independently Zermelo arrived at the same paradox, see [154], footnote 9.

Consider $a = \{x \mid x \notin x\}$)[30]), it is asked whether $a \in a$ or $a \notin a$. If $a \in a$ then a is one of the x's with the property $x \notin x$, so $a \notin a$. If on the other hand $a \notin a$, then by the definition of a we have $a \in a$.

This paradox was so elementary, in the sense that it uses no set theoretical machinery whatsoever, that its publication made everybody realize that something was wrong with set theory.

The mathematician-logician GOTTLOB FREGE was hit hardest of all. Frege had in

---

[30]) $\{x \mid \ldots x \ldots\}$ is a notation for "the set of x such that $\ldots x \ldots$".

1879 published a treatise on a symbolism for logic, the Begriffsschrift [41][31]) and on the basis of this he had built a formalization of arithmetic, laid down in the two volumes of his "Grundgesetze der Arithmetik, begriffsschriftlich abgeleitet"[32]). The second volume was ready to appear, when the author received a letter from Russell ([61], p. 124) who pointed out that Frege's system permits the derivation of the paradox of the set of all sets, that are not a member of themselves. Russell concludes in this letter that apparently not all definable collections form a totality (i.e. set). This dramatic event is recorded by Frege in the epilogue to the Grundgesetze:

> "To a scientific author hardly something worse can happen, than the destruction of the foundation of his edifice after the completion of his work. I was placed in this position by a letter of Mr. Bertrand Russell, when the printing came to a close."

Frege's problems originate in his admitting of extensions of functions (Wertverläufe von Funktionen), or, in different terminology, in admitting definable collections as sets (Comprehension, cf. p. 47). Russell's paradox made the precarious situation of set theory clear, not only to mathematicians, but also to philosophers etc., who did not have at their disposal sufficient mathematical knowledge to appreciate Cantor's or Burali-Forti's results. Many versions of the Russell paradox have been proposed. A popular one is the one concerning the village barber. In a certain village there is a barber who shaves all men, who do not shave themselves. Question: does the barber shave himself? Answer: yes and no[33]).

The reaction of the mathematical world was varied. As usual the working mathematician went on, avoiding these fancy entities like 'the set of all sets', in the hope that in due time the mess will be cleared up.

Those concerned with the foundations of mathematics however felt that action was necessary.

Dedekind, for instance, felt so discouraged by the appearance of the paradoxes, that he held back the third edition of his "Was sind und sollen die Zahlen?" for eight years until 1911 ([34], p. 343). because of "doubts concerning the correctness of important fundamentals".

The attempts to 'save' set theory can be considered as revisions of the concept of set. One way to do so was to exclude 'too large" sets, as these seemed to be present wherever paradoxes appeared in set theory.

The best-known systematic revisions of set theory are those of Russell and Zermelo.

---

[31]) Also in van Heyenoort [61] and in [43].

[32]) The basic laws of arithmetic, derived in the begriffsschrift.

[33]) There is not really a paradox here, one merely concludes that such a barber does not exist.

Russell together with A. N. Whitehead worked out a systematization of mathematics in the monumental *Principia Mathematica* (1910-1913). Russell also at one time ([119] cf. [61], p. 150) proposed 3 other solutions to the difficulties caused by the paradoxes: (i) the zigzag theory, (ii) the limitation of size theory, (iii) the no-class theory. As Russell's work is strongly based on logic, we will not discuss it here.

To Zermelo's work we will return in section 4.

3   The axiom of choice

Around the turn of the century set theory seemed to have run into a dead end, Cantor had established many wonderful results and the methodological advantages of set theory were gratifying, but all the same, important problems in the theory of cardinals and other parts of set theory withstood attempts to solve them. One such problem was the comparability of cardinals, i.e. the property that for any two cardinals m, n we have $m < n$ or $m > n$ or $m = n$.

A major breakthrough in this area was provided by ERNST ZERMELO in his proof that every set can be well-ordered [153]. The proof, which appeared in the Mathematische Annalen in 1904, was part of a letter from Zermelo to Hilbert. Zermelo wrote that the proof grew out of a conversation with ERHARD SCHMIDT. After describing the proof, Zermelo adds the following comments. The present proof rests upon

> ... "the principle that even for an infinite totality of sets there are always mappings that associate with every set one of its elements, or, expressed formally, that the product of an infinite totality of sets, each containing at least one element, itself differs from zero. This logical principle cannot, to be sure, be reduced to a still simpler one, but it is applied without hesitation everywhere in mathematical deduction."

The problem of well-ordering arbitrary sets (in particular infinite ones) was not new. Cantor already in 1883 ([25], p. 169) noticed the 'extremely remarkable law of thought with general validity' that every (well-defined) set can be well-ordered. He promises to return to this law later. Although he announced a proof several times, the problem was still open when Hilbert in 1900 listed the well-ordering problem in his famous address at the international mathematical congres in Paris [62].

The new principle that Zermelo mentioned is known by the name of *Axiom of Choice*. (Auswahlaxiom, AC for short). This principle had already been used in mathematics without being recognized. A recorded exception is provided in the case of Peano. In a paper on differential equations, [10] (1890), Peano, on the occasion of the construction of a function, had to choose values in intervals infinitely often. He did this by exhibiting a rule, remarking:

> "Mais comme on ne peut pas appliquer une infinité de fois une loi arbitraire avec laquelle à une classe *a* on fait correspondre un individu de cette classe, on a formé ici une loi determiné, avec laquelle à chaque classe *a* sous des hypothèses convenables on fait correspondre un individu de cette classe[34])."

Peano unambiguously rejects the axiom of choice here.

Also Beppo Levi (1902), criticizing Bernstein's proof that the cardinality of a partition of a set is equal to or less than the cardinality of the set, points out that a new principle is involved.

Notwithstanding these earlier apprehensions it is only fair to credit Zermelo for the axiom of choice, considering his masterly insight in the importance and applicability of the principle.

The introduction of the choice principle raised fervent discussions and disagreements among mathematicians. Already the next volume of the Mathematische Annalen contains four papers criticizing Zermelo. We will consider some of the objections to the axiom of choice, as they are characteristic for the mathematical attitude of those days and because some have survived to this very day. Zermelo, in view of the controversies his paper had stirred up, returned to the well-ordering theorem and the criticism in 1908 in his paper "Neuer Beweis für die Möglichkeit einer Wohlorderung" [154]. Peano in [103] reproached Zermelo for not proving the axiom of choice and he added that he himself had not been able to prove it in his "Formulaire" [102].

Zermelo in his answer [154] acutely dissects Peano's objections: (i) in mathematics non-provability does not imply nonvalidity (think of axioms), and (ii) Peano's Formulaire might very well be incomplete at the point of the axiom of choice (and, adds Zermelo, as there are no infallible authorities in mathematics, we must also take that possibility into account and not reject it without objective examination).

In an effective mockery, Zermelo corrects Peano's statement of the underivability of the axiom of choice in his "Formulaire". All Peano has to do, he says, is derive the Russell paradox (which is derivable in Peano's system) and thence derive the axiom, as everything can be proved from a contradiction.

As we already have seen Poincaré objected to Cantorism (set theory) as a whole (partly under the erroneous identification of logic and set theory); curiously enough he accepted the axiom of choice: "Hence this is a synthetic judgement a priori; without it the theory

---

[34]) But as one cannot apply infinitely often an arbitrary law, which associates to a class *a* an individual of that class, here a well-determined law has been formed, which makes, under suitable hypotheses, correspond to each class *a* an individual of that class.

of cardinals would be impossible, for finite numbers as well as for infinite ones" ([109], les mathématiques et la logique).

Other objections were based on the suspect character of the set W of all ordinals (an inconsistent set, to quote Cantor), those were easily refuted by Zermelo. The set W continued to bother people for a long time. Hessenberg ([60], § 98) writes: "The set W itself is, incidentally, ungrateful in the highest degree for all attempts to redeem its honor. Thus in volume 60 of the Mathematische Annalen both Bernstein and Jourdain exert themselves on behalf of its consistency, the former proving on the basis of properties of W that there are sets that cannot be well-ordered, while the latter succeeds in proving the opposite.".

The first thought of a proof of the well-ordering theorem of a new-comer would be to exhaust a set by successive choices, an idea also entertained by Cantor and Bernstein. Borel [20] also mentions the same construction, but rejects it without further comment, claiming to have reduced therefore the axiom of choice to absurdity. Of course Borel is totally besides the point, one does not prove non P by exhibiting a faulty proof procedure that does not prove P. In Borel's textbook on function theory "Leçons sur la Theorie des Fonctions" (2nd ed. and later) an extensive discussion of the axiom of choice and set theory is included.

In it one finds the opinion of a group of French mathematicians known as the semi-intuitionists. Their objections (and those of many working mathematicians) can be summarized by the statement

"The choice function (or set) is not nameable (definable in finitely many words) and hence the axiom of choice cannot be considered as a meaningful principle for the founding of mathematics[35])."

Generally speaking it is the non-constructive character of the axiom of choice that has repelled some and attracted others[36]).

The reader should, however, not think that only objections resulted from Zermelo's bold paper. A number of mathematicians straight away applied the axiom of choice in various branches of mathematics. E.g. G. Hamel showed the existence of a basis of the set of real numbers (as a vector space over the field of rationals) in [54] (1905) and G. Vitali established the existence of a set of reals that is not Lebesgue-measurable in [149] (1905) (cf. Ch. II p. 104).

The usefulness of AC in algebra is beautifully illustrated in Steinitz' book "Theorie der algebraische Körper" [142].

---

[35]) See also [14].

[36]) Lebesgue, in [78], 1907, proved the following curious result: there is no measurable choice-function for the collection of countable subsets of the continuum.

Certain strange misconceptions concerning the axiom of choice were fostered by some mathematicians (e.g. A. DENJOY and P. LEVY) who claimed that the proper version of the axiom of choice is 'a non-empty set contains at least one element'. Needless to say that the above statement is simply a tautology involving no set theory at all. As late as 1964 P. Levy published an extremely confused exposition [81] on the axiom of choice and related subjects.

The intuitionist L. E. J. BROUWER (who at that time was mainly known for his work in topology) commented in his Ph. D. Thesis [21] on the well-orderingtheorem. He declares the theorem to be downright false. His arguments, which are correct, when viewed in the light of the intuitionistic conception of mathematics, must be considered in the perspective of his later systematization. They are two in number: (i) the continuum, as a growing medium, does not consist of isolated individuals, so this prevents a well-ordering, (ii) all well-ordered sets (in the sense of Brouwer) are countable.

The history of the well-ordering theorem certainly had its dramatic episodes. One of these had JULIUS KÖNIG in the leading part. König, who had a reputation of great precision and acuteness, delivered at the Third International Congres of Mathematics in Heidelberg a lecture in which he established that the cardinality of the continuum cannot be an aleph and hence cannot be well-ordered. Cantor, who attended the congress, was deeply shocked, as he was firmly convinced of the truth of well-ordering principle.

Fortunately (for Cantor) a loophole in König's proof was soon discovered. König published his proof in the Mathematische Annalen and also indicated the loophole [71].

It is a pleasure to read König's paper, it is of great perspicuity and precision. We will sketch the contents here, as there are interesting arguments involved, and also because it showed that set theory had by then attained quite a high level.

Let $\mathfrak{m}_i$ be a cardinal ($i \in \omega$) then *König's theorem* asserts that

$$\sum_{i\in\omega} \mathfrak{m}_i \leqslant \prod_{i\in\omega} \mathfrak{m}_i \leqslant \left(\sum_{i\in\omega} \mathfrak{m}_i\right)^{\aleph_0}$$

moreover if $\mathfrak{m}_i < \mathfrak{m}_{i+1}$, then $\sum_{i\in\omega} \mathfrak{m}_i < \prod_{i\in\omega} \mathfrak{m}_i$ [37]).

---

[37] Operations on cardinals were defined by means of operations on sets, e.g.

$\sum_{i\in B} \overline{\overline{A_i}} = \overline{\overline{\cup_{i\in B} A_i}}$, where the A_i's are mutually disjoint

$\prod_{i\in B} \overline{\overline{A_i}} = \overline{\overline{\times_{i\in B} A_i}}$, where $\times$ denotes the cartesian product

$\overline{\overline{A}}^{\overline{\overline{B}}} = \overline{\overline{A^B}}$, where A^B denotes the set of all mappings of B into A

Consider the sequence $\aleph_\mu$, $\aleph_{\mu+1}$, $\aleph_{\mu+2}$, $\ldots$

then $\mathfrak{s} = \sum_{i\in\omega} \aleph_{\mu+i} < \prod_{i\in\omega} \aleph_{\mu+i} = \mathfrak{p}$.

It is clear that $\sum_{i\in\omega} \aleph_{\mu+i} = \aleph_{\mu+\omega}$.

According to König's theorem

$$\mathfrak{s} < \mathfrak{p} \leqslant \mathfrak{s}^{\aleph_0}, \text{ so } \mathfrak{s}^{\aleph_0} \leqslant \mathfrak{p}^{\aleph_0} \leqslant \mathfrak{s}^{\aleph_0^2} = \mathfrak{s}^{\aleph_0}$$

Hence $\mathfrak{p}^{\aleph_0} = \mathfrak{s}^{\aleph_0} > \mathfrak{s}$. Now suppose $2^{\aleph_0} = \aleph_{\mu+\omega} = \mathfrak{s}$, then $(2^{\aleph_0})^{\aleph_0} = \mathfrak{s}^{\aleph_0} > \mathfrak{s} = 2^{\aleph_0}$.

But it is well known that $(2^{\aleph_0})^{\aleph_0} = 2^{\aleph_0}$ (already established by Cantor), so we have obtained a contradiction. Conclusion $2^{\aleph_0} \neq \aleph_{\mu+\omega}$ for all ordinals μ[38]). So far the reasoning is impeccable and, as a matter of fact, recent results (see p. 61) show that the above inequality sums up all our knowledge of the location of the cardinality of the continuum among the cardinals.

However König went on to show that for no μ $2^{\aleph_0} = \aleph_\mu$ can hold. The proof was based on the following identity, taken from Bernstein's Ph. D. thesis: $\aleph_x^{\aleph_0} = \aleph_x \cdot 2^{\aleph_0}$. If we suppose $2^{\aleph_0} = \aleph_\mu$ and we again consider $\aleph_{\mu+\omega}$, then

$$(\aleph_{\mu+\omega})^{\aleph_0} = \aleph_{\mu+\omega} 2^{\aleph_0} = \aleph_{\mu+\omega} \cdot \aleph_\mu = \aleph_{\mu+\omega}.$$

This contradicts the inequality $\mathfrak{s} < \mathfrak{p} \leqslant \mathfrak{s}^{\aleph_0}$.

König's conclusion was: the cardinality of the continuum is not an aleph. If the continuum were well-orderable, then its cardinality would be an aleph[39]), hence the continuum is not well-orderable. It turned out however, that Bernstein's identity does not hold for all indices[40]).

In a later paper [72] König came back to the possibility of well-ordering the continuum. This time he based his refutation of the well-orderability of the continuum on the concept of definability. He reasoned as follows: consider the set E of the reals definable in finitely many words. $\bar{E} = \aleph_0$, $2^{\aleph_0} > \aleph_0$, so the complement F of E is not empty. If the continuum were well-ordered then there would be a smallest element x_0 in F. But x_0

---

[38] $2^{\aleph_0}$ is the cardinality of $\{0,1\}^N$, i.e. the set of all mappings from the natural numbers into $\{0,1\}$. This set is the well-known Cantor set, which has the same cardinality as the continuum, so $2^{\aleph_0}$ denotes the cardinal of the continuum. Also $2^{\aleph_0}$ denotes the cardinality of the powerset of N.

[39] This is immediate, because for some ordinal a we would have $2^{\aleph_0} = a$. Then the least ordinal β such that $2^{\aleph_0} = \underset{1}{\beta}$ is an aleph (and it exists because a is well-ordered).

[40] See [11], [56], and also [77] p. 287 ff.

has a finite definition. Contradiction. Conclusion: $2^{\aleph_0}$ has no well-ordering and is not an aleph.

Other paradoxes based on definability are those of Richard and Berry, these were much simpler and involved no set theory. To understand the definability paradoxes one needs considerably more formal apparatus than König possessed. For a contemporary analysis see for instance [13]. The contradiction obtained by adding König's and Zermelo's result is sometimes called the Zermelo-König paradox.

Poincaré, in a serie of talks at the university of Göttingen [108], commented on the relation between Cantor's result $2^{\aleph_0} > \aleph_0$ and the countability of the set of finitely definable reals (the only existing ones according to the semi-intuitionists, including Poincaré). His solution to the paradoxal situation is to deny an actual contradiction, which according to him is only apparent. The trouble with Richard's (and König's) definability is (according to Poincaré) that it admits impredicative definitions. E.g. the number x_0 in F was defined with respect to the set of all definable reals. A proper way to deal with definable reals would be to start with a suitable starting set M_0, e.g. the integers, from M_0 to define M_1, from $M_0 \cup M_1$ to define M_2 etc. Poincaré (following Richard [112]) now reconciles Cantor's and Richard's result, by indicating that Richard tells us that, whenever the process is broken off, there is at least one 'new' definition while Cantor tells us that the process never stops. Hence, he concludes, there is no contradiction. Actually, this does not solve the problem, as one does not get beyond the countable.

Equivalents of the axiom of choice

In a number of cases authors employing the axiom of choice have claimed that the axiom was necessary for the desired result (e.g. Zermelo did so in the case of the statement that the cardinality of a partitioning of a set S is equal to or less than the cardinality of S [153]).

For certain statements this claim has been validated by an actual proof of their equivalence to the axiom of choice. An extensive survey of such statements has been given by H. RUBIN and J. RUBIN in the monograph "Equivalents of the axiom of choice". The reader is referred to this book for more information and for proofs.

One might ask why there was (and still is) so much interest in establishing equivalences to AC. As already pointed out, many mathematicians considered AC as a hazardous principle and hence a proof not employing AC was generally prefered to one using AC So a quite normal question was: "can it also be proved without AC?". At the earlier stages of set theory two ways were open with respect to such a question: (i) one actually found a proof not using AC, (ii) one established the equivalence of the statement under consideration and AC. In both cases the status of the statement was made clear. There remained however a number of theorems that were derivable with the help of *AC*, but from which one could not derive *AC* (such as the Boolean Prime Ideal theorem). The status of these theorems, and of *AC* itself, became clear only after the independence results of Cohen. To make the notion of equivalence precise we should specify under which assumptions the equivalences hold. Most of the time the results are obtained in the system of Zermelo-Fraenkel, however in some cases stronger results can be obtained. Instead of specifying the axiomsystem we refer the reader to [117].

We will list here a few examples of equivalences.

1. The well-ordering theorem.

We have seen that Zermelo formulated the axiom of choice and exploited it to show that every set can be well-ordered. The converse (namely the implication of *AC* by the well-ordering theorem) is easily established.

Let *S* be a non-empty set of non-empty sets, consider the union *U* of all these sets and well-order *U*. Then define $f(A) = $ least element of *A*, clearly *f* is a desired choice function.

2. The multiplicative axiom.

If each $x \in A$ is non-empty then $\prod_{x \in A} x \neq \varnothing$ [41]).

The multiplicative axiom is merely a reformulation of *AC*.

3. Various maximum principles, due to Hausdorff, Kuratowski, Zorn, Teichmüller, Tukey, Moore-Bourbaki and others. Mostly these principles are indiscriminatingly called *Zorn's lemma*.

Kuratowski was the first to establish a principle of this kind (1922): A family of sets closed under union of well-ordered subfamilies (well-ordered with respect to inclusion) contains a maximal set.

We will exhibit two of these principles.

a. If *R* is a transitive antisymmetric binary relation on a non-empty set *A* and each *R*-chain has an upper bound, then there exists an *R*-maximal element[42]).

---

[41]) $\varnothing$ denotes the empty set.

[42]) If *R* is a binary transitive relation then an *R*-chain is a set linearly ordered by *R*. An *R*-maximal element *a* has the property that for all *x* aRx implies $a = x$.

b.   If $A \neq \emptyset$ and if the union of each $\subseteq$-chain is an element of A, then A has a $\subseteq$-maximal element.

4.   Algebraic principles.

a.   Every proper ideal in a lattice with a 1 is contained in a maximal ideal (D. Scott, 1954 [126]).

b.   If B is a Boolean algebra and $S \subseteq B$, such that $0 \notin S$ then there is a maximal ideal $I \subseteq B$ such that $I \cap S = \emptyset$ (S. Mrowska, 1955).

c.   Every group contains a maximal abelian subgroup (Klimovsky 1962 [69]).

Actually Klimovsky's theorem differs slightly from the above statement, the reader should consult the papers. Felgner has obtained a very short proof for Klimovsky's theorem and at the same time generalized it, e.g.: $AC \Rightarrow$ Every group G contains maximal metabelian subgroups. (It should be mentioned that the axiom of Foundation (cf. p. 49) is not used in Klimovsky's or Felgner's proof.)

5.   Cardinal principles.

$\mathfrak{m}$, $\mathfrak{n}$, $\mathfrak{p}$ are infinite cardinals.

a. $\mathfrak{m} + \mathfrak{n} = \mathfrak{m} \cdot \mathfrak{n}$ (Tarski, 1924)

b. $\mathfrak{m} = \mathfrak{m}^2$ (Tarski, 1924)

c. $\mathfrak{m} + \mathfrak{p} < \mathfrak{n} + \mathfrak{p} \Rightarrow \mathfrak{m} < \mathfrak{n}$ (Tarski, 1924)

d. $\mathfrak{m} < \mathfrak{p}$ and $\mathfrak{n} < \mathfrak{p} \Rightarrow \mathfrak{m} \cdot \mathfrak{n} < \mathfrak{p}$ (Sierpinski, 1946)

e. $\mathfrak{m} + \mathfrak{n} = \mathfrak{m}$ or $\mathfrak{m} + \mathfrak{n} = \mathfrak{n}$ (Lesniewski, Sierpinski, 1947)

f. $\mathfrak{m} < \mathfrak{n}$ or $\mathfrak{n} < \mathfrak{m}$ or $\mathfrak{m} = \mathfrak{n}$ (Hartogs, 1915)

6.   Tychonoff's theorem.

The cartesian product of compact topological T_1-spaces is compact. Tychonoff proved $AC \Rightarrow$ *Tychonoff's theorem* in 1935, Kelly proved the converse in 1950.

7.   The theorem of Skolem-Löwenheim (R. Vaught, 1956).

This is a theorem from logic, we will return to it later.

8.   If for all $x \in A$ $f(x) \underset{1}{<} g(x)$ then $\bigcup_{x \in A} f(x) \underset{1}{<} \prod_{x \in A} g(x)$.   (König 1895, Zermelo 1908).

9. **(NBG)**[43]). If X is a proper class (cf. p. 47) and x a set, then $x \underset{1}{<} X$ (Von Neumann) (i.e. x can be properly embedded in X).

10. **(NBG)**. If X is a proper class and x a set, then X can be mapped onto x.

There are also weaker forms of AC and principles which are implied by AC, but which seem not to imply it. The actual establishment of the non-derivability of AC from other principles had to wait until after 1963. Some examples are:

The axiom of countable choice (AC^ω). If A is a countable set and all its elements are non-empty then there exists a choice function f such that $f(x) \in x$ for all $x \in A$.

The axiom of dependent choices (DC). Let R be a binary relation on A, such that for each x there exists a y with $R(x,y)$ then there is a function f on the natural numbers such that $R(f(i), f(i+1))$, for all natural numbers i.

The axiom of dependent choices was first introduced by Bernays in 1942 ([7], p. 86), he showed that $AC \Rightarrow DC \Rightarrow AC^\omega$. Later Tarski independently introduced DC.

The ordering principle. Each set can be ordered.

The Boolean prime ideal theorem (BPI). Every proper ideal in a Boolean algebra is contained in a prime ideal. *BPI* is equivalent with the completeness theorem for first order predicate logic.

BPI is sufficient to prove the Hahn-Banach theorem, although most textbooks simply apply Zorn's lemma.

The relation between these and other principles will be discussed in section 7.

4   Zermelo takes over

Set theory was rapidly developing into an independent, accepted branch of mathematics. Especially in analysis and topology set theoretic methods found a ready appreciation. The acceptance as a legitimate mathematical discipline was underlined by the publication of the first textbooks on set theory. Already in 1900 A. SCHOENFLIES published in the Jahresberichte der Deutsche Mathematiker Vereinigung, Vol. 8, part 2 an extensive survey of set theory under the title "Die Entwicklung der Lehre von den Punktmannig-

---

[43])   **NBG** will be introduced on p. 45.

ERNST ZERMELO

faltigkeiten"[44]). In 1913 an enlarged second edition appeared as a seperate book "All-gemeine Theorie der unendliche Mengen und Theorie der Punktmengen". In England W. H. YOUNG and G. C. YOUNG published in 1906 their book "The theory of sets of points".

In 1914 F. HAUSDORFF's "Grundzüge der Mengenlehre" appeared. This excellent textbook did not find its match for many years, and even today it is worth to consult the Grundzüge der Mengenlehre on quite a number of subjects. The first of a long series of texts on set theory by Fraenkel appeared in 1919, even in 1972 a new edition of one of his books appeared.

Criticism, some well-founded and some ill-founded, had not been spared on the young discipline. Notably from the side of the constructivists objections were raised against the generosity of set theory with respect to the existence of sets. Poincaré, for instance, doubted the existence of the second number class[45]), Brouwer went beyond doubting and plainly denied the existence of the second number class[46]).

These objections, added to the consternation caused by the paradoxes, made people realize that sets were not as easily described as Bolzano, Dedekind or Cantor did. Gradually mathematicians, remembering the success of Hilbert's Grundlagen der Geometrie, started to look for an axiomsystem.

---

[44]   The development of the theory of pointmanifolds.

[45]   [108], p. 48.

[46]   This is one of his Stellingen (Theses), which in Holland accompany a Ph. D. thesis [21].

Hessenberg in 1906 [60] writes:

"Until now nobody has succeeded in defining correctly the notion of set. Probably we cannot expect a definition, but rather an axiomsystem. The usual definitions of set do not allow any useful conclusions, but also they allow paradoxical sets. ... But as there seem to be consistent, infinite sets, a suitable definition or a correct axiom system should exclude paradoxal entities."

Of course one must except the constructivist mathematicians here, as they more or less vaguely defined their notion of set. A precise definition of set (Menge), but tailored for intuitionistic purposes, was given by Brouwer [23].

The historic paper that opened the way for modern set theory was Zermelo's paper *"Untersuchungen über die Grundlagen der Mengenlehre I[47])"*, which appeared in 1908 [155]. Zermelo's axiomatization is not formalized as we find it today in textbooks, it is rather an axiomatization in the spirit of Hilbert's Grundlagen der Geometrie.

In Zermelo's theory there are two kinds of entities: urelements[48]) and sets.

Sets are distinguished from urelements by the property that they possess elements (with the exception of the empty set $\varnothing$). Because the theory was not formulated in a formal language, Zermelo wanted to make precise what kind of statements are significant for set theory; he defined a statement G to be *definite* if the fundamental relations of the domain, by means of the axioms and the universally valid laws of logic, determine without arbitrariness whether it holds or not. It is clear what Zermelo meant by the above description, but it tends to put the responsibility on (the unspecified) logic. A rigorous formalization would in due time ban the vagueness.

As a matter of fact Cantor experienced the same qualms. In 1880 he calls a manifold (set) of elements, belonging to some conceptual domain, well-defined, if on the basis of its definition and the logical principal of the excluded third, it is internly determined, (i) whether some object from the same conceptual domain belongs to the manifold or not, and (ii) whether any two objects from the set, in spite of formal differences in their mode of being given, are identical or not. All this flood of words served to make the notion of 'set' clearer, but it is doubtful whether it did. It showed that the intensional aspects of mathematics, which set theory so succesfully was to ban, were fully realised by Cantor (and others). Fraenkel and Skolem were to take up the notion of 'definiteness' again (cf. p. 39, 42).

In the theory of Zermelo there were the following primitive notions: set, urelement,

---

47) Investigations in the foundations of set theory.
48) Urelemente in German. In English literature there is no generally accepted translation of the term, so we will follow current usage and employ the term 'urelement'.

∈-relation. The axioms of Zermelo listed below, are strongly influenced by Dedekind's Erklärungen (definitions) from "Was sind und sollen die Zahlen?".

We will write down the axioms and at the same time give a formulation in symbolic notation[49]).

The existence of urelements brings certain complications along with it, for example the axiom of extensionality does not have the well-known simple form

$$(\forall z)(z \in x \leftrightarrow z \in y) \to x = y$$

because this would force us to identify any two urelements. Namely an urelement x has no elements, so $z \in x \leftrightarrow z \in y$ is vacuously true and therefore we would have $x = y$ for any two urelements.

Zermelo avoids this pitfal by stipulating in I that x and y are sets. The formal counterpart of this proviso is to introduce a primitive predicate $\mathfrak{M}(x)$ for 'x is a set'.

Zermelo defines a set as an object that contains another object, and he adds to the family of sets the *empty* (or *null*) set as an exception. In effect he calls the empty set fictitious.

I   *Axiom of extensionality* (Axiom der Bestimmtheit).
If every element of the set x is an element of the set y and vice versa, then $x = y$[50]).

$$(\forall x)(\forall y)(\mathfrak{M}(x) \land \mathfrak{M}(y) \land \forall z(z \in x \leftrightarrow z \in y) \to x = y).$$

---

[49])   A short dictionary for the symbols is given below.
If $A, B, \ldots$ are statements, then

$A \lor B$	stands for "A or B"
$A \land B$	stands for "A and B"
$A \to B$	stands for "if A, then B"
$\neg A$	stands for "non A"
$A \leftrightarrow B$	stands for "A if and only if B, or A iff B"
$\forall x\, A(x)$	stands for "for all x $A(x)$"
$\exists x\, A(x)$	stands for "there exists an x such that $A(x)$"
$\{x \mid A(x)\}$	stands for "the collection of all x such that $A(x)$"
$x \subseteq y$	stands for "x is a subset of y"

In $\forall x\, A(x)$ (or $\exists x\, A(x)$) the variable is bound. Otherwise x is called free. A variable can occur both free and bound in a formula, but one can always avoid this by replacing the bound (occurrences of) variables by new variables.
If K is a set of formula's then $K \vdash A$ stands for "A is derivable from K".
Models will be denoted by Gothic capitals and $\mathfrak{A} \models A$ stands for "A is valid (true) in $\mathfrak{A}$".

[50])   Zermelo expresses the axiom in the following condensed version: every set is determined by its elements.

II   *Axiom of elementary sets* (Axiom der Elementarmengen).
(i)   there is a (fictitious) set, that contains no element, i.e. the empty set $\varnothing$ [51]).
(ii)   If x is an object then there exists a set $\{x\}$ containing only x ($\{x\}$ is called 'single-ton x').
(iii)   If x and y are objects then there exists a set $\{x,y\}$ containing precisely x and y.

(i) $(\exists x)(\forall y)(y \notin x)$
(ii) $(\forall x)(\exists y)(\forall z)(z \in y \leftrightarrow z = x)$
(iii) $(\forall x)(\forall y)(\exists z)(\forall u)(u \in z \leftrightarrow (u = x \lor u = y))$

Note that in the formalized versions of (i), (ii), (iii) no use of braces is made. Actually one can only introduce a new term if it is clear that it denotes a unique object. Fortunately the axiom of extensionality implies the uniqueness of 'the empty set', 'the singleton', 'the unordered pair', so that the use of $\varnothing$, $\{x\}$ and $\{x,y\}$ is justified [52]). Strictly speaking the use of these terms in the axioms is unjustified.

III   *Axiom of separation* (Axiom der Aussonderung).
If $A(x)$ is definite for all substitutions of (names of) elements of z then z contains a subset containing precisely those elements x of z for which $A(x)$ holds.

$(\exists y)(\forall x)[x \in y \leftrightarrow (x \in z \land A(x))]$ (restriction: y may not occur free in $A(x)$).

Note that III supplies us with infinitely many axioms, namely for each formula $A(x)$ there is a corresponding Aussonderungs axiom. Such an expression is called an *axiom-schema*.

IV   *Axiom of the powerset* (Axiom der Potenzmenge).
If x is a set, then there exists a set $\mathscr{P}(x)$ containing precisely the subsets of x.

$(\forall x)(\exists y)(\forall z)(z \in y \leftrightarrow z \subseteq x)$.

V   *Axiom of the union* (Axiom der Vereinigung).
If x is a set then there exists a set $\bigcup x$ which contains precisely all elements of elements of x.

$(\forall x)(\exists y)(\forall z)(z \in y \leftrightarrow (\exists u(z \in u \in x))$ [53]).

---

[51]   Zermelo uses 0 instead of $\varnothing$.
[52]   This so-called *extension by definition* is treated in logic, e.g. see Schoenfield [131], p. 57.
[53] $z \in u \in x$ is shorthand for $z \in u \land u \in x$.

VI   *Axiom of choice* (Axiom der Auswahl).

If x is a set of disjoint non-empty sets then there is a set y such, that the intersections of y with the elements of x contain each exactly one element.

A precise symbolic form would take up too much space, so that we will later present one for a more suitable version of *AC*.

VII   *Axiom of infinity* (Axiom des Unendlichen).

There exists a set that contains the empty set and with each element y contains $\{y\}$

$$(\exists x)[\emptyset \in x \wedge \mathfrak{W}(x) \wedge (\forall y)(y \in x \rightarrow \{y\} \in x)].$$

We will follow current mathematical practice and refer to the axioms in abbreviated form, e.g. we will say "... holds by Infinity" instead of "... holds by the axiom of Infinity". The formulation and notation of the axioms does not follow Zermelo, we have adopted a convenient notation (resp. formulation) whenever possible.

The axiom of infinity is actually a mathematical version of Dedekind's argument for the existence of infinite sets. It is a considerable step forward by Zermelo to recognize the need to postulate an infinite set.

The axiom of choice is formulated without the use of functions. The axioms that guarantee the existence of *'larger'* sets are the Axiom of union and the Powerset axiom (apart from *AC*), but they do not involve too large sets. Hence it is plausible that sets like Ω cannot occur. For example the 'set of all sets' (say V) can be defined by $\{x \mid x = x\}$, but this is not an instance of 'Aussonderung'. Moreover V cannot be reached by iteration of the powerset operation, or the unionoperation. So it is highly improbable that V can be granted the status of set. Zermelo's axiom system limits the size of the possible sets by not allowing axioms that entail the existence of Cantor's inconsistent (very large) sets. The axiom of seperation (which we will call 'Aussonderung' for short) does only 'introduce' smaller sets.

As Zermelo does not have ordered pairs at his disposal, he has to define functions (mappings) in a roundabout way, he does so in an ingenious way. The paper contains the principal properties of the notions involved and the theorem of König (see p. 28) and a theorem concerning two notions of infinity.

The notion of definiteness, as we already said, lacks precision with Zermelo. The main reason for introducing this notion was the Aussonderung axiom, one had to make precise what kind of properties defined subsets. In itself, the vague notion was good enough for axiomatizing naive set theory and for obtaining all known results, but it was certainly not sufficient for sophisticated problems like the independence of the axiom of choice.

This last problem was succesfully attacked by Fraenkel in 1922 [36]. It is not surprising that Fraenkel offered a revised version of the Aussonderung axiom. He defines

D. van Dalen

ABRAHAM A. FRAENKEL*

a function to be a rule of the following kind: "an object $\varphi(x)$ shall be formed from a ('variable') object x that can range over the elements of a set, and possibly from further given ('constant') objects, by means of a prescribed application (repeated only a finite number of times, of course, and denoted by φ) of Axioms II, IV, V"[54]).

Now the Aussonderung axiom reads: If M and two functions φ and ψ are given then M contains subsets $M' = \{x \in M \mid \varphi(x) \in \psi(x)\}$ and $M'' = \{x \in M \mid \varphi(x) \notin \psi(x)\}$.

Fraenkel's independence proof for the axiom of choice is an outstanding achievement, showing that the potential possibilities of axiomatic set theory were fully realized. The method of proof employed by Fraenkel is modelled after a popular illustration of AC given by Russell [119] in 1905:

If $\aleph_0$ pairs of boots are given then we can select from each pair a boot by a choice function, e.g. for each pair takes the left boot. But when we are given $\aleph_0$ pairs of socks, then no such choice function can be specified.

Fraenkel exploits this idea by constructing a model (in modern terminology) with a collection of urelements $a_1, \bar{a}_1, a_2, \bar{a}_2, a_3, \bar{a}_3, \ldots$ The sets $\{a_i, \bar{a}_i\}$ are called cells (the 'pairs of socks'). The proof rests on the fundamental lemma:

for each set M there exists an $A_M \subseteq A = \{\{a_1, \bar{a}_1\}, \{a_2, \bar{a}_2\}, \ldots\}$ such that (i) A_M is co-finite[55]) and (ii) M is invariant if the elements in the cells of A_M are permuted.

Suppose now that C is a *choice set* of A (i.e. C meets every set in A in exactly one element). Determine A_C and permute one pair in A_C, then A and C are invariant,

---

* Photo by Ricarda Schwerin, Studio Alfred Bernheim, Jerusalem

[54]) i.e. $\varphi(x)$ is obtained by repeatedly constructing singletons, pairs, power sets and unions from x and given sets.

[55]) i.e. $A - A_M$ is finite.

but, because of the permutation, C cannot be invariant. Contradiction. Hence AC does not hold, so AC is not provable from the other axioms.

The construction of the model sets the pattern for all later work. The idea is simple enough, first throw in all elements that have to be in the model (here the a_i, $\bar{a}_i$'s, the cells, $A \oslash$ and the natural numbers) and then apply all the operations that the axioms prescribe[56]). E.g. form pairs, unions, powersets, subsets according to Aussonderung. For the resulting model one has to prove the axioms of course.

The method of Fraenkel fits theories with urelements, because in these one can permute urelements without worrying too much about the consequences for other sets, i.e. the 'structure' of the model is only slightly changed, no axioms are violated. It is considerably harder to work out similar procedures for theories without urelements. The independence of AC in its general setting was not established by Fraenkel, and it remained unsolved until 1963, when a completely new approach was found.

Fraenkel's method has been taken up by a number of mathematicians: Mostowski [90], Lindenbaum, Bernays, Specker, Mendelssohn et al. Some of them eliminated the use of urelements by using sets, which violate the axiom of Foundation (see p. 49), in particular sets x with the property that $x = \{x\}$, so-called *reflexive sets* (cf. [141]).

The models, obtained by the above methods are called permutationmodels. The method using urelements is called the *Fraenkel-Mostowski method*, the one without urelements is called the *Fraenkel-Mostowski-Specker-method* (*FMS*-method for short). Quite an amount of group theory is applied in the modern permutation models, on the whole the *FMS*-method embodies an attractive part of mathematics. For more information see the monograph [35] of U. Felgner.

In the same year (1922) a Scandinavian mathematician, THORALF SKOLEM, presented a lecture on axiomatized set theory [134] to the Fifth Congress of Scandinavian Mathematicians in Helsinki. Skolem already had won his spurs as a logician by his work on satisfiability of logical formulas. His name is mostly known for the Skolem-Löwenheim theorem, but he also contributed to widely varying mathematical disciplines as number theory, recursion theory, set theory and model theory[57]).

In the above mentioned lecture he considers a number of points that are of considerable interest for set theory (and logic). We will consider some of them.

(i) Independent of Fraenkel (and others probably) Skolem proposes an improved formulation of the Aussonderung axiom.

---

[56]) Fraenkel apparently thought this procedure so obvious, that he spends only one sentence on it ([61], p. 287).

[57]) For more information on Skolem the reader is referred to the contributions of Fenstad and Hao Wang to the 'Selected Works in Logic by Skolem' [137].

He defines a definite expression as one built up from (atomic) propositions of the form $a \in b$ and $a = b$, by means of the logical connectives (see p. 37). In modern terminology: a definite expression is a formula of first order predicate calculus with primitive relation symbols '$\in$' and '$=$'. Skolem's proposal ultimately coincides with Fraenkel's, but it is an improvement in the sense that it is easier to handle and to visualize. Eventually Skolem's notion was adopted in axiomatic set theory.

The formulation of Aussonderung now reads: If $A(x)$ is a formula of set theory[58]) then $(\forall x)(\exists y)(\forall z)(z \in y \leftrightarrow (A(z) \wedge z \in x))$. Zermelo later (apparently unaware of Skolem's proposal) gives analogous specifications of 'definite' [155], see also [135] and [39], p. 40. But he rejects the proposals of Fraenkel and Skolem because they rest on the concept of natural number, which set theory should not presuppose. He therefore introduces an axiomatic treatment of 'definiteness', which, however, never has found acceptance.

(ii) The theorem of Skolem-Löwenheim[59]), when applied to a model of set theory, asserts that if set theory has any model at all, it has a countable model. Skolem describes this phenomenon by the term 'relativity'. That is, the concepts of set theory in the axiomatic setting are relative with respect to the model. For example Cantor's theorem tells us that there are uncountable sets, so in the countable model mentioned above there is an uncountable set. This state of affairs is sometimes called the *paradox of Skolem*. Skolem himself, in [134], remarks that there is no paradox at all, a set may be uncountable inside a model (while countable from the outside), because in the model there is no function, which maps the set one-one onto the set of integers (in the model!).

The above considerations make it clear that set theory cannot be characterized in an absolute way by an axiomatization[60]).

(iii) Another defect of Zermelo's axiom system was noted by Skolem; it turns out that certain, generally excepted sets cannot be proved to exist in Zermelo's system **Z**. At approximately the same time Fraenkel and Mirimanov noticed the defect [37], [86].

---

[58]) i.e. set theory formalized in first order predicate logic.

[59]) The theorem of Skolem-Löwenheim, in the version required here, states that if a first order theory has an infinite model, then it has a countable model (provided the language of the theory is at most countable).

We distinguish nowadays two versions of the Skolem-Löwenheim theorem: If the theory T has a model $\mathfrak{A}$ of cardinality $\mathfrak{m}$ (in symbols $\|\mathfrak{A}\| = \mathfrak{m}$) and $\mathfrak{m} > \mathfrak{n}$ where $\mathfrak{n}$ is the cardinality of the language of T, then T has a model of cardinality $\mathfrak{n}$.

(ii) the upward Skolem-Löwenheim theorem: If T has a model $\mathfrak{A}$ with $\|\mathfrak{A}\|$ greater than the cardinality of the language, then for every $\mathfrak{m} > \|\mathfrak{A}\|$, T has a model of cardinality $\mathfrak{m}$. For more information see [131], [68], [5]. Actually, there are more refined versions, using the notion of elementary substructure.

[60]) And not only set theory, but (almost) every axiomatic theory, e.g. arithmetic, cf. Skolem's fundamental paper [136].

THORALF SKOLEM*

Skolem and Fraenkel proposed remedies. Skolem shows by a simple (model-theoretic) argument, that in **Z** it is not provable that $\{M,\ \mathscr{P}(M),\ \mathscr{P}^2(M),\ \mathscr{P}^3(M),\ldots\}^{51})$ is a set.

We will give his argument in modern terminology on p. 51. Skolem proposed a new axiom of the following form: let U have two free variables, such that $(\forall x \in a)(\exists ! y)\, U(x,y)$, where a is a set, then $\{y \mid (\exists x \in a)(\, U(x,y)\}$ is a set too.

A managable version is: If the domain of a function is a set then the range is a set too. Fraenkel came up with the same formulation, although Fraenkel's notion of function (see p. 40) is somewhat awkward.

The new axiom is called the axiom of replacement (Ersetzungs Axiom) and it reads in symbols:

$$(\forall x)(\exists ! y)\, A(x,y) \to (\forall u)(\exists v)\big[y \in v \leftrightarrow (\exists x)(x \in u \wedge A(x,y))\big]$$

Already Cantor recognized, to a certain degree, the fact, that images of sets must be sets. In his letter to Dedekind of 28-7-1899 ([25], p. 444 or [61], p. 113) he writes "Two equivalent multitudes either are both 'sets' or both inconsistent".

(iv) The last remark of Skolem concerns AC.

"As long as we are on purely axiomatic grounds", he says, "there is nothing

---

* Reproduced from: The Theory of Models. Proceedings of the 1963 International Symposium at Berkeley. Edited by J. W. Addison, Leon Henkin and Alfred Tarski. North-Holland Publishing Company, Amsterdam 1965.

[61]) $\mathscr{P}(M)$ is the power set and $\mathscr{P}^{n+1}(M) = \mathscr{P}(\mathscr{P}^n(M))$.

wrong with *AC*. But most mathematicians do not take the axiomatic view. We can ask: What does it mean for a set to exist if it can perhaps never be defined? It seems clear that this existence can only be a manner of speaking, which can lead only to purely formal propositions – perhaps made up of very beautiful words – about objects called *sets*. But most mathematicians want mathematics to deal, ultimately, with performable computing operations and not to consist of formal propositions about objects called this or that."

In the meantime an important concept had been defined by N. Wiener, F. Hausdorff and K. Kuratowski, namely the *ordered pair*. The definition of a pair is easy to give: the pair $\{a,b\}$ is the set having a and b as its only elements, in symbols

$$c = \{a,b\} \leftrightarrow (\forall x)(x \in c \leftrightarrow x = a \lor x = b),$$

Kuratowski's definition of ordered pair is now universally excepted, it reads:

$$<a,b> = \{\{a\}, \{a,b\}\}.$$

One easily checks that $<a,b> = <c,d>$ iff $a = c$ and $b = d$.

It is important to have the notion of ordered pair available in set theory, as it allows a very simple definition of relations and, above all, functions. We can define a function as a set of pairs such that to each element in the domain there exists exactly one element in the range. To be precise, we can define a predicate Fnc, Fnc(f) means „f is a function" by the following: $\mathrm{Fnc}(f) \underset{d}{=} (\forall x)(\forall y)(\forall z)[(<x,y> \in f \land <x,z> \in f) \rightarrow y = z]$.

The domain of a function f is defined by

$$\mathrm{Dom}\,(f) = \{x \mid (\exists y)(<x,y> \in f)\}$$

and likewise the range

$$\mathrm{Range}\,(f) = \{y \mid (\exists x)(<x,y> \in f)\}.$$

The above definition of function goes back to Peano (1911) [104]. The ordered pair also enables one to define (cartesian) products:

$$x \times y = \{<u,v>, \mid u \in x \land v \in y\}.$$

Ordered triples, quadruples etc. are defined by iteration of the formation of ordered pairs:

$$<x,y,z> = <<x,y>, z>$$
$$<x,y,z,w> = <<x,y,z>, w>.$$

In this section we have tried to sketch the import of Zermelo's ideas and the consequences of his bold initiatives. In spite of his outstanding achievements, his place in the scientific community seems to have been modest. Zermelo was born in Berlin in 1871, he obtained his 'Habilitation' in 1899 in Göttingen and in 1910 he obtained a professorship in Zürich.Then from 1916-1926 he was a private scholar (Privatgelehrte) in Germany, until he was made honorary professor in Freiburg (Germany). Not being of a yielding nature, Zermelo got into trouble with the Third Reich authorities (e.g. he refused to bring the 'Hitlergruss') and he decided to resign before being dismissed. In 1946 he was rehabilitated. His death in 1956 was relatively unobserved (see [106]).

5   Making inconsistent sets respectable

Due to the efforts of Zermelo, Fraenkel and Skolem an axiom system which was Zermelo's system **Z**, improved by replacing the Aussonderungs axiom by the Replacement axiom, was gradually accepted by the mathematical world. Historically this system bears the names of Zermelo and Fraenkel, it is denoted **ZF°**. In mathematical practice **ZF°** has however been superseded by the system **ZF**, obtained from **ZF°** by adding the axiom of foundation (see p. 49). It is the system **ZF**, that nowadays is referred to as the theory of Zermelo-Fraenkel. In order not to confuse the reader, we will from now on discuss **ZF** and mention **ZF°** only, when the absence of the axiom of foundation is essential.

The system can be viewed, as an attempt to codify the practice of set theory, while avoiding too large 'sets', such as the set of all ordinals (limitation of size).

In 1925 a young man of 22 years old, arrived at the scene and presented a novel axiom system in which there was place for ghosts, like 'the set of all sets'.

Already two years earlier this novice, JOHANN VON NEUMANN, had published a paper "On the introduction of transfinite numbers" [95] in which he defined ordinals not, as was done in the naive phase of set theory, as equivalence classes of well-ordered sets, but by singling out specific sets.

According to Von Neumann, a set α is an ordinal, if (i) it is well-ordered by inclusion, (ii) every element of α is the union of all its predecessors. One immediately realizes that $\varnothing$ fits the definition and that if α is an ordinal, so is $\alpha \cup \{\alpha\}$. In particular we can build now the 'finite' ordinals

$$0 = \varnothing$$
$$1 = \{\varnothing\}$$
$$2 = \{\varnothing, \{\varnothing\}\}$$
$$3 = \{\varnothing \{\varnothing\}, \{\varnothing, \{\varnothing\}\}$$
etc.

One also sees that $\{0, 1, 2, 3, \ldots, n, \ldots\}$ is again an ordinal, it is Cantor's ω.

Von Neumann develops as much of the theory of ordinals, as is required to show that the usual properties hold. The advantages of the new approach over the older definition is twofold. In the first place the cumbersome equivalence classes and ordered sets are avoided, in the second place an elementary definition is provided that only uses the $\in$-relation. R. M. Robinson presented a simplified definition of ordinals in [114]: an ordinal α is a set such that (i) if $x \in \alpha$ and $y \in x$ then $y \in \alpha$, (ii) if $x \in \alpha$ and $y \in \alpha$ then $x \in y$ or $y \in x$ or $x = y$. The property of α mentioned in (i) is called *transitivity*, (ii) requires α to be linearly ordered by $\in$.

We will just list some properties of ordinals for future references (for proofs the reader is referred to any textbook, e.g. [77]). Ordinals are denoted by α, β, γ, ...
1. The *successor* $\alpha + 1$ of an ordinal is defined as $\alpha \cup \{\alpha\}$.
2. Ordinals which are not successors are called *limit ordinals*, e.g. ω is a limit ordinal.
3. The ordering relation of the ordinals is the $\in$-relation. We will often denote it by $<$.
4. $(\forall \alpha)(\forall \beta)(\alpha < \beta \vee \beta < \alpha \vee \alpha = \beta)$.
5. Every non-empty set of ordinals has a least element.
6. $[A(0) \wedge (\forall \alpha)(\forall \beta < \alpha) A(\beta) \rightarrow A(\alpha)] \rightarrow (\forall \alpha) A(\alpha)$.
   The so-called *Principle of transfinite induction*.

   The principle of transfinite induction is already implicit in Cantor's Beiträge ([25], p. 336 ff.). Hessenberg gives an explicit formulation in [60], 1906.

A useful version of the principle is the following:
   If $A(0)$ and $(\forall \alpha)(A(\alpha) \rightarrow A(\alpha + 1))$ and $(\forall \alpha < \lambda) A(\alpha) \rightarrow A(\lambda)$ for all limit ordinals λ then $(\forall \alpha) A(\alpha)$.

Parallel to this *proof by induction* there is a *definition by recursion*, of which we will indicate a special case.
7. If g and h are functions and a is a set then there exists a unique function f defined on the ordinals such that

$$\begin{cases} f(0) = a \\ f(\alpha + 1) = g(f(\alpha)) \\ f(\lambda) = h(\{f(\alpha) \mid \alpha < \lambda\}). \end{cases}$$

Examples of definition by transfinite recursion occur already with Cantor.

In his paper "An axiomatization of set theory" [99] Von Neumann boldly admits the things that were considered 'inconsistent sets' by Cantor. The trick consists of distinguishing between entities that can be members of other entities and those that cannot. The first correspond to sets and the second to 'inconsistent sets'. Von Neumann's paper does not make easy reading as it is formulated in terms of functions rather than sets. P. Bernays, K. Gödel, R. M. Robinson and others have given Von Neumann's system

John von Neumann*

a more manageable form. We will follow the latter presentations in which the basic ideas are better recognized. In the literature the system is mainly known under the name **NBG** (if the axiom of foundation is included), after Von Neumann, Bernays and Gödel. The system **NBG** has as primitive terms *class* and the $\in$-*relation*. Classes X, such that $X \in Y$ for some Y are called *sets*. The remaining classes are called *proper classes*[62]).

Gödel incorporates in the system (see [49]) two primitive terms $\mathfrak{Cls}$ and $\mathfrak{M}$ for the following purpose:

$\mathfrak{Cls}\ X$ stands for 'X is a class' and $\mathfrak{M}\ X$ stands for 'X is a set'. In order to facilitate working with the theory, we will follow the customary practice to denote sets by $a, b, c \ldots x, y, z$ and classes by $A, B, C \ldots X, Y, Z$ (we consider a 'two-sorted language'). Comparing **NBG** to **ZF** one can say that in **ZF** one only talks about sets and properties of sets, in **NBG** these properties determine concrete objects, i.e. classes.

An advantage of **NBG** over **ZF** is the so-called *finite axiomatizability*. In **ZF** the axiom of replacement is actually a set of axioms, namely for each formula $A(x)$ one, therefore **ZF** has infinitely many axioms. In **NBG** on the other hand finitely many axioms are sufficient, compare for example Gödel's monograph [49]. As a matter of fact the following *comprehension principle* is provable:

$$(\exists X)[(\forall x)(x \in X \leftrightarrow A(x))]$$

if $A(x)$ contains no bound class-variables.

The comprehension principle asserts that appropriate conditions define classes. E.g.

---

*   Reprinted with permission from Mrs. Neuman-Eckart. Reprint from: John von Neuman, Collected Works. Edited by A. H. Taub. Pergamon Press Ltd., Oxford 1961.

[62])   A distinction between sets and classes can be found already in König's paper [72].

$\Omega = \{\alpha \mid \text{Ord}(\alpha)\}$, where $\text{Ord}(\alpha)$ stands for the formula that expresses 'α is an ordinal'; $V = \{x \mid x = x\}$, V is the universal class.

One may wonder what happens if the restriction on the bound variables is lifted. The result is an impredicative schema (classes are introduced by reference to all classes!) that essentially strengthens the theory. For example one can prove in the resulting system the consistency of **ZF**, a fact that cannot be proved in **ZF** itself (at least if **ZF** is consistent) according to a famous theorem of Gödel[63]).

The system with impredicative comprehension schema was introduced by Morse, as an appendix in Kelly's textbook on topology [67] (see also Morse [89]). Because of Mostowski's contributions to the theory of the above system, it is nowadays called the system **M** of Morse-Mostowski. **M** is a very convenient theory to use as a meta-theory of **ZF** (see [92]).

In the last section of the paper [96] Von Neumann claims that it is consistent to add an axiom (VI 4) banning infinitely descending sequences $\ldots \in a_3 \in a_2 \in a_1$.

Mirimanoff already in 1917 [86] discussed the possibility of extraordinary sets (ensembles extraordinaires), consisting of these descending sequences. Fraenkel also considered these sets ([37] and [38], p. 355) and he suggested to eliminate them by a special 'Axiom der Beschränktheit' (axiom of limitation), stating that there are no sets but those required by the axioms of **ZF°**. Von Neumann quite correctly objects to this axiom on the grounds that it either relies on naive set theory, or that it requires a 'higher order' set theory. But then it is not clear that the axiom is valid (cf. [61], p. 404).

There is one specific axiom of Von Neumann's system that is no longer used in **NBG**, it was in a way an extension of Cantor's statement "equivalent multitudes are either both 'sets' or both inconsistent".

This axiom F reads: "A class is a proper class if and only if there is a one-one mapping from V (the universal class) onto it".

In symbols $Pr(X) \leftrightarrow X \underset{1}{=} V$ (where $Pr(X)$ stands for 'X is a proper class').

Von Neumann showed that F can be derived from the usual axioms plus a strong version UC of AC.

UC is the *axiom of universal choice*, and it states that there exists a choice function for all non-empty sets (i.e. $V - \{\varnothing\}$):

$$(\exists F)(Fnc(F) \wedge (\forall x)(x \neq \varnothing \rightarrow F(x) \in x)).$$

Also the axiom F implies UC and Replacement. It is easy to see that UC holds, just consider Ω (the class of all ordinals). By mimicking the Burali-Forti paradox, we see

---

[63]) See e.g. [131] p. 213.

that Ω is a proper class, hence $V = \Omega$. Therefore the universe is well-ordered. Now it follows immediately that there is a universal choice function.

Where von Neumann in 1925 explicitly forbids sets with descending $\in$-sequences (i.e. sequences of sets $x_0, x_1, x_2, x_3, \ldots$ with the property $\ldots x_{n+1} \in x_n \in \ldots \in x_2 \in x_1 \in x_0$), in 1929 [98] he introduces an axiom, that turned out to make life comfortable for set theoreticists. It is the so-called *axiom of foundation* (Foundation for short)[64].

The axiom of foundation reads: Every non-empty class has an $\in$-least element.

In symbols: $(\forall X)(X \neq \varnothing \rightarrow (\exists x)(x \in X \land x \cap X = \varnothing))$.

The axiom was independently introduced by Zermelo in [157] 1930. Von Neumann showed in [98] that Foundation can consistently be added to the other axioms, i.e. if the other axioms are consistent then the system obtained by adding Foundation is consistent too.

We will sketch the importance of Foundation, using essentially modeltheoretic methods.

Let us define a mapping R by transfinite recursion

$$\begin{cases} R(0) = \varnothing \\ R(\alpha + 1) = \mathscr{P}(R(\alpha)) \cup R(\alpha) \\ R(\lambda) = \bigcup_{\alpha < \lambda} R(\alpha) \end{cases}$$

(this device goes back to von Neumann).

Using Foundation, we can prove that R exhausts the universe, i.e. $(\forall x)(\exists \alpha)(x \in R(\alpha))$. Suppose not, then there is a set x such that x belongs to no $R(\alpha)$, therefore $A = \{x \mid \neg (\exists \alpha)(x \in R(\alpha))\} \neq \varnothing$. Now we apply Foundation: let x_0 be the $\in$-least element of A.

$x_0 \neq \varnothing$, because $\varnothing \in R(1)$, so x_0 belongs to no $R(\alpha)$, but all elements y of x_0 belong to some $R(\beta_y)$. By taking the supremum γ of the set of β_y's (which is possible because of Replacement), we find that all elements of x_0 are elements of $R(\gamma)$, i.e. $x_0 \subseteq R(\gamma)$. But then $x_0 \in R(\gamma + 1)$. Contradiction.

The axiom of Foundation allows us to cut up the universe in slices. The universe is often picturesquely represented as a kind of funnel. At the tip we start with $\varnothing$ and then it grows exponentially. This 'cutting up' of the universe gives more insight in the structure of it (see fig. 2a).

E.g. one can define a 'rank' of sets as the least ordinal α such that the set occurs in $R(\alpha + 1)$, to be precise $\rho(x) = \alpha$, iff $x \in R(\alpha + 1)$ but not $x \in R(\alpha)$, this definition goes back to Mirimanoff (1917, [86]). The $R(\alpha)$'s are useful in many ways, e.g. they can be used to show that certain classes are sets, all one has to do is to estimate the rank of the set.

---

[64]) Also called *axiom of regularity*, *axiom of restriction*, *axiom of groundedness*.

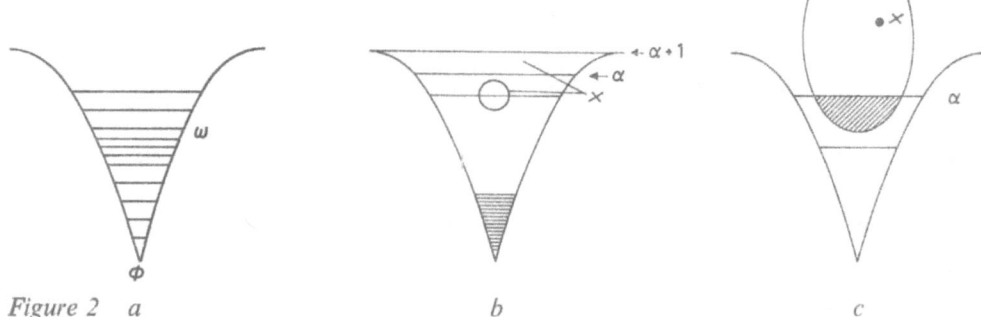

Figure 2   a   b   c

In particular if one can find an upper bound α for the ranks of all elements of x, then $x \in R(\alpha + 1)$ and hence x is a set. (see fig. 2b).

Dana Scott used the notion of rank to save the Frege-Russell cardinal [127]. The notion of cardinal, with Frege and Russell, was introduced as the equivalence class under the equivalence relation $\underset{1}{=}$. Unfortunately these cardinals are *proper classes*, as is easily demonstrated in the case of the cardinal of a singleton (e.g. $\{\varnothing\}$). Clearly all singletons are in one-one correspondence, hence the cardinal of $\{\varnothing\}$ (notation $\overline{\overline{\{\varnothing\}}}$) contains all singletons. But the class S of all singletons is in one-one correspondence with V (by $F(x) = \{x\}$), so S is a proper class.

As it is for several reasons highly undesirable to have proper classes for cardinals (for example cardinals would not exist as objects in **ZF**), it would be good to have a notion of cardinal, such that cardinals are sets.

Scott presented such notion by not taking all sets equivalent to a given set, but only those of minimal rank. In the picture this is represented by the shaded part (fig. 2c). In symbols:

$$\overline{\overline{x}} = \{y \mid y \underset{1}{=} x \wedge (\forall z)(\rho(z) < \rho(y) \to \neg \, (z \underset{1}{=} x))\}.$$

Likewise one can 'save' ordertypes and other notions introduced by equivalence relations.

Von Neumann's proof that the axiom of Foundation can consistently be added to the other axioms is one using *inner models* [65]. Suppose a model $\mathfrak{M}$ of the axioms of set theory, with Foundation deleted, is given. Then inside $\mathfrak{M}$ a model $\mathfrak{N}$ can be constructed such that all the axioms, including Foundation, hold.

One first remarks that ordinals can be defined, and their properties be proved, in a

---

[65]   The name is introduced by Sheperdson in [130].

50

system without Foundation (e.g. see [84]). So one can consider the ordinals inside $\mathfrak{M}$ and perform the construction of the $R(\alpha)$'s. In particular $\mathfrak{M}$, seen from the outside, is a set (in the real world, or some informal set theory), therefore there is a first ordinal ρ which is not in $\mathfrak{M}$ (the supremum of all ordinals in $\mathfrak{M}$)[66]). One can show that $R(\rho)$ is a model of all the axioms including Foundation.

Now suppose the axiom system including Foundation was inconsistent, then it could not have a model. But then the axiom system could not have a model either, by the above argument, so it was already inconsistent. This concludes the proof.

The independence of Foundation of the remaining axioms was shown by Bernays in [7] 1954 (but the result was already announced by him in 1941 ([7], p. 10) and by Specker in 1951 in his Habilitationsschrift (cf. [141])[67]).

The system $\mathbf{ZF}^0$ supplemented with Foundation is called $\mathbf{ZF}$, formulations can be found in most textbooks on set theory (e.g. [30], [77], [39], [131]).

Zermelo in his paper [157] also considers models of set theory of the form $R(\rho)$ (for suitable ρ). He includes urelements so his construction is a slight generalization of the one given above; start with $R(0) =$ the set of all urelements and the empty set. Zermelo derives a kind of normal form for models of set theory, depending on the set of urelements and the ordinal ρ (i.e. two models are isomorphic if and only if their sets of urelements are equivalent and their ordinals ρ and ρ' are identical). The models $R(\rho)$ and the ordinals ρ are called Normalbereiche (normal domains) and Grenzzahlen (boundary numbers) by him. The sets $R(\alpha)$ can also be used to establish the independence of the axiom of infinity. One easily shows that $R(\omega)$ is a model for all axioms of $\mathbf{ZF}$ except Infinity. An ingenious elementary proof of the independence of Infinity was put forward by Ackermann [1] (1937): consider the set of natural numbers ω and define on ω an $\in$-relation by $x \in y$ if $y/2^x$ is odd, put $\varnothing = 0$. It is a matter of calculation to show that the axioms of $\mathbf{ZF}$ hold except for Infinity.

Also the independence of Replacement of $\mathbf{Z}$ can be shown in a similar way (Bernays 1948 [7]). $R(\omega \cdot 2)$ is a model for $\mathbf{Z}$, but replacement fails in $R(\omega \cdot 2)$.

For example we show that the axiom of powerset holds: if $x \in R(\omega \cdot 2)$ then $x \in R(\alpha)$ for some $\alpha < \omega \cdot 2$, but then $x \subseteq R(\alpha)$ (because $R(\alpha)$ is transitive) and hence

$\mathscr{P}(x) \subseteq R(\alpha+1)$. Therefore $\mathscr{P}(x) \in R(\alpha+2)$. As $\alpha+2 < \omega \cdot 2$ we have $\mathscr{P}(x) \in R(\omega \cdot 2)$.

Replacement fails in $R(\omega \cdot 2)$ as can be seen of the 'function' f with $f(n) = \omega + n$[68]).

---

[66]) One should realize that the notion of ordinal (à la von Neuman) is *absolute* in the sense that something is an ordinal inside the model if and only if it is so from the outside.

[67]) See also Rieger [113], Hajek [53], Boffa [15], [16].

[68]) Properly speaking f is not a set in the model, so we really should consider the formula
$x \in \omega \wedge y = x + \omega$.

Paul Bernays

The 'domain' of f is ω, but the 'range' is a subset of $R(\omega \cdot 2)$ and not a subset of any earlier $R(\alpha)$, hence the range of f is not an element of the model.

The inadequacy of Aussonderung was already pointed out by Skolem and Fraenkel (cf. p. 42).

During the thirties Cantorian axiomatic set theory had assumed a more or less definite shape. The two major formalizations were **ZF** and **NBG**. The first theory flourished under the constant attention of Abraham Fraenkel, who made this new discipline accessible to general mathematicians in a number of well written books and expositions. His first exposition, "Einleitung in die Mengenlehre" (1919), was conceived and written while the author served in the army during the first world war.

The creator of the second theory, John von Neumann was an extremely versatile mathematician. Set theory was one of the many subjects that challenged his genius. The theory **NBG** no doubt owes its development and its popularity to Paul Bernays and Kurt Gödel, in particular Gödel's monograph [49] was instrumental in making the system easily accessible.

6 The consistency of the axiom of choice and the continuum hypothesis

Set theory entered a new era with the publications of K. Gödel in the years 1938-1940.

Already a considerable amount of work had been done in set theory and both the foundations and the technical aspects of set theory were in an impressive state.

Building on von Neumann's work Bernays [7] and others had brought it into a form that was easier to manage than von Neumann's original formulation in terms of I and II objects.

All the same, some fundamental problems were still unsolved, among them the famous continuum problem: does $2^{\aleph_0} = \aleph_1$ hold? (and the generalized continuum problem: does for all α $2^{\aleph_\alpha} = \aleph_{\alpha+1}$ hold?). The identity was already conjectured by Cantor in 1878. Initially Cantor was quite convinced of the validity of the continuum hypothesis, in 1883 he concludes a paper ([25], p. 244) with the words "from this, with the help of the theorems proved in no. 5 §13, one can conclude that the linear continuum has the cardinality of the second number class". However in his letter of 14-11-1884 to Mittag-Leffler he writes that, renewing his attack at the continuum problem, he was startled:

"When I was toiling these days, . . ., what did I find? I found an *exact* proof that the continuum does not have the cardinality of the second number class and moreover that it does not have any (cardinal-) number[69]) as its cardinality."

The next day he sents off a letter to supersede the first one:

"My dear friend,
The grounds about which I wrote you yesterday, against the theorem of the second cardinality of the linear continuum, were today again *refuted*; therefore all grounds that the continuum has the second cardinality again force, undefeated, itself to the front and I hope finally to decide soon this most important problem once and for all either positively or negatively . . ." ([122], p. 18).

Apart from fundamental objections to the continuum hypothesis (*CH*), as harbored by the French semi-intuitionists and Brouwer, several attempts were undertaken to refute *CH*, we already mentioned the erroneous exposition of König (p. 29).

Hilbert recognized the importance of *CH* already in 1900 when he made it the first of his famous list of mathematical problems. He also tried to solve *CH* as appears from his paper "Über das Unendliche" [63] (1926) (see also [61]).
Hilbert considers the Baire space ω^ω, instead of the set of real numbers and he tries to find a mapping from ω_q (second number class) onto ω^ω. This work considerably differs from usual set theoretic procedures, therefore this survey is not the right place to

---

[69]) i.e. aleph.

comment on it. It be sufficient to remark that Hilbert's paper gave rise to the study of hierarchies of number theoretic functions and the association of ordinals to different types of recursive functions. Zermelo, in 1928, remarked to Paul Levy that nobody in Germany understood what Hilbert had in mind in this paper ([81], p. 89).

Gödel, in a letter (1965) to van Heyenoort (cf. [61], p. 368, 369) wrote that there is a remote analogy between Hilbert's work and his. Hilbert only used constructive operations and transfinite recursion with respect to constructive ordinals, while Gödel admits arbitrary ordinals and recursion. Gödel's notion of 'constructible' is a broad kind of constructivity, e.g. ordinals are constructible, but not all ordinals are constructive. Many authors considered the continuum hypothesis and no doubt many editors of journals were annoyed by fallacious proofs. It is not customary for mathematicians to vote on open issues (or to conduct polls), so we don't have records of the opinion of the mathematical world on *GCH*. In some instances we happen to know a private opinion, so Lusin conjectured in 1935 that *GCH* might be independent of **ZF** (in particular that $2^{\aleph_0} = 2^{\aleph_1}$ would be consistent). So, until Gödel's papers appeared, the continuum problem was wide open. Gödel did not settle the dispute, but he at least showed that *GCH* and *AC* were harmless principles, in the sense that $\mathbf{ZF} + GCH + AC$ is consistent if **ZF** is. Therefore the assumption of *GCH* cannot by itself lead to contradictions, and also $\neg GCH$ cannot be proved from **ZF** (assuming of course that **ZF** is consistent). In [47] and [48] Gödel used **ZF**, but in the monograph [49] he used the version of von Neumann's system now known as **NBG**.

The approach in [48] is somewhat more intuitive then the one in [49], so we will stick to **ZF** for the time being for expounding Gödel's idea's. If one wants to show $2^{\aleph_0} = \aleph_1$, then a one-one mapping from $2^{\aleph_0}$ (the continuum) onto $\aleph_1$ is required. Now $\aleph_1$ is (in the sense of von Neumann) the first ordinal with a cardinality greater than $\aleph_0$, so the structure of $\aleph_1$ is fairly decent. As for the continuum (the powerset of ω if you like), we have not much insight in its structure. One way to look for a one-one mapping from $2^{\aleph_0}$ onto $\aleph_1$ is to *make* $2^{\aleph_0}$ nice. Gödel achieved this by constructing $2^{\aleph_0}$, and not only that but the whole universe, starting from the ordinals. Gödel's constructionprocess resembles the one using *R*, but the latter is highly inconstructive and impredicative. Instead of iterating the powerset operation, Gödel, at each step, adjoins only *definable* subsets.

We will now sketch the procedure. The system **ZF** contains the following axioms (cf. p. 37)

I	Extensionality	V	Powerset
II	Empty set	VI	Infinity
III	Pairing	VII	Replacement
IV	Union	VIII	Foundation (cf. [30], p. 50).

Suppose a model $\mathfrak{U}$, being a pair $<U,\in>$ where U is a *set* and $\in$ a binary relation in U, of **ZF** is given, that is: all axioms of **ZF** hold in $\mathfrak{U}$.

If φ is a formula then we denote '$\mathfrak{U}$ satisfies φ' by $\mathfrak{U} \vDash \varphi$.

If a is a set in U then we say that b is *definiable from* a (in $\mathfrak{U}$) if

$$b = \{x \mid \mathfrak{U} \vDash \varphi_a(x, a_1, \ldots, a_n), a_1, \ldots, a_n \in a\}$$

for some formula φ_a.

φ_a is the formula obtained from φ by restricting all variables to a (i.e. replace $\forall x (\exists x)$ by $\forall x \in a (\exists x \in a)$).

Example: for each element c in a transitive set a $\{c\}$ is definable from a by the formula

$$(\forall y \in a)(y \in x \leftrightarrow (\forall z \in a)(z \in y \leftrightarrow z \in c)).$$

Now we define a mapping on the ordinals of $\mathfrak{U}$ by

$$\begin{cases} M(0) & = \varnothing \\ M(\alpha+1) = (M(\alpha))'^{70}) \\ M(\lambda) & = \bigcup_{\alpha < \lambda} M(\alpha), \text{ where } x' \text{ is the set of all sets definable from } x. \end{cases}$$

If ρ is the supremum of all ordinals in $\mathfrak{U}$, then the elements of $L = M(\rho)$ are called the *constructible sets* (in $\mathfrak{U}$).

The property of a set to be constructible can be formulated in **ZF** by a formula $L(x)$: x is constructible $\Leftrightarrow \mathfrak{U} \vDash L(x)$ (cf. footnote 49). Therefore we can formulate in **ZF** the axiom.

"All sets are constructible" $((\forall x) L(x)$ or, as is customary, $V = L$). The axiom $V = L$ is called the axiom of constructibility. There is a definable well-ordering of the constructible sets, this is fairly plausible, because the universe of constructible sets (L) is split up in levels parallel to the ordinals and at each level the order of elements can be determined by means of their defining formulae. Hence it follows that $L \vDash AC$. Gödel showed that $L \vDash \mathbf{ZF} + AC + GCH + V = L$. In particular $L \vDash V = L$ shows that from the point of view of L all sets are constructible, so an iteration of the construction process, described above, does not lead to a proper submodel.

From the above modeltheoretic considerations one can (by means of the completeness theorem for first order theories) obtain the following:

(1)   **ZF** is consistent $\Rightarrow$

       ZF has a model $\Rightarrow$

---

[70]) Note that for finite ordinals i $M(i) = \mathscr{P}(i) = R(i)$.

$ZF + AC + GCH + V = L$ has a model $\Rightarrow$

(2) $ZF + AC + GCH + V = L$ is consistent.

Actually, Gödel's presentation of the consistency results, although motivated by interpretations, is of a prooftheoretic nature. That is to say, independent of any interpretation one can, by just sticking to the notion of proof, show that if a contradiction can be derived e.g. from $ZF + V = L$, it can also be derived from ZF alone.

Gödel's work was carried on by Sheperdson [130], who systematized and extended the results.

The axiom of constructability, introduced by Gödel, has a curious status. Although it is useful for technical work, it is not very plausible and Gödel himself does not favour a set theory based on $V = L$. As $V = L$ is a rather strong condition in the universe, the axiom has proved to be useful e.g. in hierarchy theory (cf. [2]).

GEORG KREISEL has pointed out a remarkable application of Gödel's construction of the inner model of constructible sets: If an arithmetical formula is provable from $ZF + V = L$, then it is already provable from ZF (cf. [35], p. 41). Since from $V = L$ both GCH and AC are provable one also obtains: If an arithmetical formula is provable from $ZF + GCH + AC$ then it is already provable from ZF. In this way one can eliminate the use of the continuum hypothesis and the axiom of choice.

We will now list some results and problems in various areas of set theory.

The consistency of ZF (and NBG) still is an open problem. From the second theorem of Gödel it follows that the consistency of ZF cannot be proved in ZF. Hence a consistency proof will have to use pretty strong principles and it is problematic what good that will do to the status of set theory. On the other hand it is still possible that someone comes up with a specific proof that shows ZF inconsistent. With respect to consistency ZF and NBG are equally strong, that is ZF is consistent if and only if NBG is. There is a very simple proof for this equivalence based on model theory (see [30], p. 77). Actually it is a corollary of the following fact. If φ is a sentence in the language of ZF (contains no class variables or terms) then $ZF \vdash \varphi \Leftrightarrow NBG \vdash \varphi$[71]). A distinction between ZF and NBG can be made on the basis of the formal system. NBG as we already mentioned has a finite set of axioms (cf. [49], [30], [84]), ZF has an infinite set of axioms (remember that Replacement is actually an axiom schema, i.e. it stands for infinitely many axioms), but even stronger, Montague has shown that no finite set of axioms is sufficient to derive all axioms (and hence all theorems) of ZF [88].

The relation between GCH and AC was determined by Tarski and Lindenbaum in a joint paper in 1926 [143] where they stated that AC is a consequence of GCH. The proof in the literature was provided by Sierpinski (see also Specker [140]).

---

[71] $\Gamma \vdash \varphi$ (where Γ is a set of formulas) stands for "φ is derivable from Γ".

A proof is also presented in Cohen [30], p. 148 ff.

We have expressed so far the continuum hypothesis in terms of alephs, this is by no means necessary. An aleph-free formulation is $\mathfrak{m} < \mathfrak{n} \leqslant 2\mathfrak{m} \Rightarrow \mathfrak{n} = 2\mathfrak{m}$.

CH has a number of remarkable consequences, in Sierpinski's monograph "Hypothèse du Continu" (1934) an extensive list of these is exhibited. We will present a few here.

C 49   There exists a plane set E such that each vertical section has measure 0 and also each horizontal section of $\bar{E}$[72]).

C 55   Every comeager linear set contains uncountable many disjoint comeager sets.

C 62   There exists a real function which is discontinuous on every uncountable domain.

P 10   In the Hilbert space there exists an uncountable set having no uncountable finite dimensional subsets.

P 2   The plane is the union of countably many curves[73]).

One can say that *CH* is not very evident because there are no plausible statements that imply *CH*, while many plausible statements imply $\neg$ *CH* (cf. Gödel [50]).

The axiom of foundation, not having the same degree of evidence as the other axioms, has been the object of research. It is easily seen that $\mathbf{ZF}^0 + \text{Foundation} \vdash NIDC$[74]).

On the other hand $\mathbf{ZF}^0 + AC + NIDC \vdash \text{Foundation}$. Mendelson in [83] showed that AC is necessary by establishing $\mathbf{NBG} + NIDC \nvdash \text{Foundation}$[75]).

7   The independence of the continuum hypothesis

Gödel's results on the relative consistency of AC and GCH did not settle the continuum problem. In order to solve the continuum problem one way or another one, either had to derive GCH in one of the formal theories (say $\mathbf{ZF}$), or to provide a countermodel. Mathematicians had grown sceptic as to the possibility of a proof of GCH, they rather hoped for a countermodel.

Church in his lecture at the International Congress of Mathematics in 1966 at Moscow

---

[72])   A vertical (horizontal) section of E is the intersection of A and a vertical (horizontal) line, $\bar{E}$ is the complement of E.

[73])   A curve is a set of the form $\{< x, y > \mid y = f(x)\}$ or $\{< x, y > \mid x = g(y)\}$.

[74])   NIDC stands for the axiom "there are no infinite descending chains of the form $\ldots \in a_3 \in a_2 \in a_1$".

[75]) $\ldots \nvdash \ldots$ stands for "not $\ldots \vdash \ldots$"

D. van Dalen

KURT GÖDEL*

[10] related that Gödel already in 1942 had found an independence proof for $V = L$ (by way of a theory of types), this method could probably be extended to an independence proof of AC. Again, in 1946, Gödel returned to the same subject. In his "Remarks before the Princeton bicentennial Conference on the problems in mathematics" (published in Martin Davis' collection [32] (1965)) he introduced the notion "definability in terms of ordinals". This concept is wider than the definability employed in the theory of constructible sets. Gödel expects that the new concept will provide a simpler relative consistency proof for AC. Myhill and Scott, independent from Gödel discovered the properties of the above notion. E. L. Post also had reached similar results in the fifties, but not published them. The results of Myhill and Scott were presented in 1967 in Los Angelos, they are contained in the proceedings. Also Takeuti and Vopenka rediscovered the notion.

As Gödel showed that $V = L \rightarrow GCH$ (the axiom of constructibility implies the generalized continuum hypothesis), one had to look for a model violating the axiom of constructability (for short: a model of $V \neq L$). Here a remarkable obstacle was met, as noted by Sheperdson [130] and Cohen [28] (cf. [30], p. 107). Sheperdson (and later Cohen) established the existence of a *minimal model* $\mathfrak{M}$, in the sense that $\mathfrak{M}$ is isomorphically contained in every (standard!) model. In this model the axiom of constructability holds (immediate from Gödel's construction). The minimal model is used to establish

* Reproduced from: Foundations of Mathematics. Symposium Papers Commemorating the Sixtieth Birthday of Kurt Gödel. Managing Editor J. J. Bulloff; Technical Editors: T. C. Holyoke and S. W. Hahn, first edition. Copyright (c) 1969 by Springer Verlag, Berlin-Heidelberg-New York.

the following fact. There is no formula $A(x)$ in **ZF** such that the relativized versions of the axioms of **ZF** and of $V \neq L$ with respect to $A(x)$ are provable in **ZF**.

Suppose the contrary and let $\mathfrak{M}$ be the minimal model of **ZF**, then the set $\mathfrak{M}'$ of all elements of M for which $A(x)$ holds (in $\mathfrak{M}$) is a model of **ZF** and $V \neq L$. But $\mathfrak{M}$ is minimal, so $\mathfrak{M} = \mathfrak{M}'$. This contradicts the fact that $\mathfrak{M} \vDash V = L$. Hence no formula $A(x)$ of the desired kind exists.

The situation strikingly differs from that in the case of $V = L$; Gödel found a formula $L(x)$ that 'defined' the constructable sets, therefore in any model $\mathfrak{M}$ he could isolate a submodel $\mathfrak{N}$ for $V = L$ (the method of *inner models*). Apparently such a method does not work for $V \neq L$ and a fortiori not for AC and GCH.

A brilliant new method was invented in 1963 by a newcomer in the field: PAUL J. COHEN. In two papers in the Proceedings of the National Academy of Sciences of the United States [29] he established the independence of the continuum hypothesis.

The idea (in terms of models) was to take a model of **ZF** and to add to it a non constructible set (as in algebra one adjoins an element to a field). The trouble is that the sets in a model are strongly interwoven, such that the addition of an extra set disturbs the model to the effect that axioms of **ZF** may turn invalid. It is interesting to compare the situation e.g. with field theory. There the 'structure' of the new element is completely irrelevant, one just introduces an object without caring how it is made up of parts. The same holds for the addition of urelements by the Fraenkel-Mostowski method. However in well-founded models of **ZF** the interior $\in$-structure of new sets matters very much. Cohen found a solution to this problem, he introduced *generic* sets which disturb as little as possible. The basic technique is that of *forcing*. Without going into details we will try to indicate the notion of forcing. A forcing condition P is a finite set of formulas of the form $n \in a$ and $n \notin a$ such that never $n \in a$ and $n \notin a$ both occur in P. One can think of the forcing conditions as finite quantities of information. The notion 'P forces the formula A' (notation $P \Vdash A$) is defined inductively:

$P \Vdash c \in d$ if $(c \in d)$ is an element of P

$P \Vdash A \wedge B$ if $P \Vdash A$ and $P \Vdash B$

$P \Vdash A \vee B$ if $P \Vdash A$ or $P \Vdash B$

$P \Vdash A \rightarrow B$ if for every condition Q with $Q \supseteq P$ $Q \Vdash A$ implies $Q \Vdash B$

$P \Vdash \neg A$ if no condition Q extending P forces A (not $Q \Vdash A$)

$P \Vdash \exists x A(x)$ if there exists a constant c such that $P \Vdash A(c)$

$P \Vdash \forall x A(x)$ if $P \Vdash \neg \exists x \neg A(x)$.

An infinite sequence of conditions $P_0 \subseteq P_1 \subseteq P_2 \subseteq \ldots$ can be constructed such that for each sentence A there is a P_n with $P_n \Vdash A$ or $P_n \Vdash \neg A$. Such a sequence is called complete. A complete sequence forces for each k (natural number) $k \in a$ or $k \notin a$, so with a a subset of ω is determined. Such a set is called *generic*. The notion of forcing provides a substitute for truth, in a way truth is approximated by finite quantities of information. If one builds, in the way Gödel did, a model containing one or more generic sets then a sentence A is true in the model if and only if some P_n of the corresponding complete sequence forces A. Hence truth in the model can be expressed in **ZF** by a suitable formalization of the notion of forcing (all these finite sequences of expressions can be coded in a suitable way). The finiteness of the forcing conditions is exploited in most proofs by permuting the model such that the forcing condition does not change but truth does, thence the invalidity of the desired statement follows.

At present many expositions of Cohen's methods have appeared, for instanc those of Cohen [30], Mostowski [92], Jensen [66], Shoenfield [131], ch. 9, Felgner [35]. The reader is referred to those texts as a detailed examination is beyond the scope of this survey.

Cohen obtained the first important results:

Suppose **ZF** has a model (even a standard model), then

(1) $\mathbf{ZF} + V \neq L + AC + GCH$ has a model
(2) $\mathbf{ZF} + AC + \neg GCH$ has a model
(3) $\mathbf{ZF} + \neg AC$ has a model
(4) $\mathbf{ZF} + \neg AC^{\omega}$[76]) has a model

All these results can be formulated in another way; e.g. (2) could be replaced by:

(2′)   If **ZF** is consistent, so is $\mathbf{ZF} + AC + \neg GCH$
       or
(2″)   If **ZF** is consistent, then $\mathbf{ZF} \nvdash AC \rightarrow GCH$.

In particular Cohen established that in the diagram

$$V = L \rightarrow GCH \rightarrow AC$$

the arrows cannot be reversed.

From the work of Gödel and Cohen we conclude that

(i)   If **ZF** is consistent, then neither AC nor $\neg AC$ can be deduced from **ZF** (i.e. AC is independent from **ZF**)
(ii)   If **ZF** is consistent, then neither GCH nor $\neg GCH$ can be deduced from $\mathbf{ZF} + AC$ (i.e. GCH is independent of $\mathbf{ZF} + AC$).

Note that the above independence results are only obtained with respect to a specific formal system. With respect to the truth according to the Platonist views of foundations the continuum problem is still open.

---

[76]) AC^{ω} is the axiom of countable choice (see p. 34).

In the wake of Cohen's work a wealth of new results was produced. We will list a number of results.

(5)   If **ZF** is consistent, so is $\mathbf{ZF} + \neg AC^{\omega} +$ there exists a Dedekind finite set that is not finite.

(5′) $\mathbf{ZF} \not\vdash x$ is Dedekind finite $\to x$ is finite. (Cohen, [30].)

Remember that according to Dedekind [33] a set is finite if it is not equivalent to a proper subset of itself. One easily proves that each finite set (i.e. equivalent to a natural number) is finite in Dedekind's sense. For the converse one has to use the axiom of choice. (5′) confirms the suspicion that one essentially needs AC.

(6)   If **ZF** is consistent, so is $\mathbf{ZF} + AC^{\omega} + \neg DC$ (Jensen).

(6′) $\mathbf{ZF} \not\vdash AC^{\omega} \to DC$.

Hence the axiom of dependent choices is properly stronger than the axiom of countable choice (cf. p. 34).

(7)   The ordering principle is independent of **ZF** (Feferman). The ordering principle reads: "Every set can be totally ordered" (cf. p. 34).

(8)   If **ZF** is consistent, so is $\mathbf{ZF} + BPI + \neg AC$.

(8′) $\mathbf{ZF} \not\vdash BPI \to AC$ (Halpern, Laüchli, Levy).

BPI is the Boolean prime ideal theorem (cf. p. 34).

(9)   If **NBG** is consistent, so is $\mathbf{NBG} + GCH + \neg UC$ (Easton)

In particular this shows that the axiom of universal choice is not a consequence of the axiom of choice.

Souslin [139] formulated the following hypothesis: Every complete total ordering without endpoints, which satisfies the countable chain condition on open intervals is order-isomorphic to the open interval (0.1).

Let us denote this hypothesis by SH.

(10₁)   If **ZF** is consistent, so is $\mathbf{ZF} + GCH + \neg SH$ (Tennenbaum, Jech).

(10₂)   If **ZF** is consistent, so is $\mathbf{ZF} + AC + SH + \neg GCH$ (Solovay, Tennenbaum).

(10′) SH is independent with respect to **ZF**.

König already in 1904 had proved that $2^{\aleph_0} \neq \aleph_{\mu + \omega}$ (see p. 30). Cohen and Solovay showed (cf. [139]) that König's result was the best possible:

(11)   If $\aleph$ is an infinite cardinal with $\aleph_0 < \mathrm{cf.}\,(\aleph)$ then $\mathbf{ZF} + 2^{\aleph_0} = \aleph$ is consistent, if **ZF** is[77]), e.g. $\mathbf{ZF} + 2^{\aleph_0} = \aleph_{27}$ is consistent, if **ZF** is.

Solovay extended this result to the effect that the generalized continuum hypothesis may hold up to a certain cardinal and be violated for this cardinal.

---

[77])   The cofinality of $\aleph$, cf. $(\aleph)$, is the least ordinal a such that $\aleph$ is the limit of an increasing a-sequence of ordinals (less than $\aleph$). It can be shown that $\aleph_0 < \mathrm{cf.}\,(\aleph_{\mu+1})$.

(12)   Vitali (p. 28) showed that $\mathbf{ZF} + AC \vdash$'not all sets of reals are Lebesgue measurable', this result has been complemented by showing that $\mathbf{ZF} + \neg AC +$ 'not all sets of reals are Lebesgue measurable' is consistent if $\mathbf{ZF}$ is. Hence the existence of non Lebesgue measurable sets of reals is not equivalent to the axiom of choice. Another result by Solovay (1965) states that $\mathbf{ZF} + \neg AC +$ 'all sets of reals are Lebesgue measurable' is consistent, if $\mathbf{ZF} +$ 'there exists an inaccessible cardinal' $+ AC$ is consistent. Apparently there is no a priori reason to believe in the existence of sets which are not Lebesgue measurable.

There are scores of new consistency results, independence results, underivability results. The reader is referred to Mathias' survey in the Proceedings of the UCLA Set Theory Institute.

After Cohen enlarged the arsenal of set theoretic methods with the forcing technique several extensions and generalizations were introduced by various mathematicians.
At present we have at our disposal forcing à la Cohen, à la Sacks, à la Solovay, à la Silver. Each of these have their specific merits, unfortunately they are beyond the scope of this exposition.
     In the meantime, in Chechoslovakia, a school in the foundations of set theory emerged. Authors like Vopenka, Hajek, Jech, Bukovský, Hrbáček contributed important results to many sophisticated areas of set theory.

It must be noted that the notion of forcing as introduced by Cohen has peculiar intuitionistic traits. There are semantics for intuitionistic logic, as provided by Beth [12], [13], Kripke [74], that closely resemble forcing. The connection between these subjects was noted by Kripke [74] and Grzegorczyk [52].

Shortly after various forms of forcing came to attention. Scott and Solovay noticed that a unified approach could be made through models of set theory with truth values in Boolean algebras. Scott [129] presented in 1966 an extremely elegant and short proof of the independence of the continuum hypothesis. An exposition of Boolean-valued models for set theory by Scott and Solovay has been announced. It is extremely gratifying that the general theory of algebraic logic has found such striking applications. A detailed presentation of the method of Boolean valued models is given by Rosser [116]; Jensen [66] also devotes a chapter to the subject.

8   Large cardinals

Gödel in his paper "What is Cantor's Continuum Problem?" [50], [6], points out that the independence of *GCH* from the axiomsystem of Zermelo-Fraenkel is not the final word on the continuum hypothesis. One might, by better insight into the notion of set, be led to add new plausible axioms to **ZF**. He mentions axioms, asserting the existence of large cardinals, which can be viewed as very strong axioms of infinity.

The axiom of infinity we considered on p. 39 is an example, it asserts the existence of a sufficiently large set. On the basis of the axiom of infinity we can prove the existence of lots of cardinals (and ordinals of course) that were not available in the system without the axiom of infinity.

We will consider some of the notions of 'large cardinal' that are in use in the literature.

8.1   *Inaccessible cardinals*

Cardinals are initials ordinals, i.e. ordinals α with the property that for all $\beta < \alpha$ $\bar{\beta} \neq \bar{\alpha}$ (after Von Neumann [97]).

A cardinal α is *singular* if it is cofinal with an ordinal β less than α (i.e. if there is a sequence $<\sigma_\gamma>_{\gamma < \beta}$, with $\sigma_\gamma < \alpha$, such that $\alpha = \lim_{\gamma < \beta} \sigma_\gamma$. A cardinal is *regular* if it is not singular. One can arrange all cardinals in a well-ordered sequence $\omega_0, \omega_1, \omega_2, \ldots, \omega_\omega, \omega_{\omega + 1}, \ldots$ (note the analogy with the alephs). Using *AC* it is easy to check that $\omega_{\alpha + 1}$ is regular. For example ω_ω is singular.

ω_α is called *weakly inaccessible* if ω_α is regular, and α is a limit ordinal. The notion of 'weakly inaccessible' was introduced by Hausdorff [57] (1908). A cardinal κ is called *strongly inaccessible* if it is regular and if for every $\beta < \kappa$ we have $2^\beta < \kappa$. This notion was introduced by Tarski [144]. Zermelo had already defined the notion of boundary number (cf. p. 51):

α is a boundary number iff $R(\alpha) \vDash$ **ZF**.

The strongly inaccessible cardinals have an analogous property: α is strongly inaccessible iff $R(\alpha + 1) \vDash$ **NBG** (as shown by Sheperdson), however the classes of strongly inaccessible cardinals and boundary numbers do not coincide.

The existence of strongly inaccessible cardinals cannot be proved in **NBG** (nor **ZF**), because otherwise the consistency of **NBG** could be shown (via $R(\alpha + 1) \vDash$ **NBG**) in **NBG** and that would, under the provision that **NBG** is consistent, contradict Gödel's second theorem.

Mahlo [82] introduced a class of yet larger cardinals in 1911. Consider the following definition: $\aleph_\alpha$ is of Mahlo-class 0 if it is strongly inaccessible, $\aleph_\alpha$ is of Mahlo-class $\beta + 1$

if there are $\aleph_\alpha$ cardinals of Mahlo-class β smaller than $\aleph_\alpha$, $\aleph_\alpha$ is of Mahlo-class λ (for limit or ordinal λ) if it is of Mahlo-class β for every $\beta < \lambda$. And, as if this were not enough, Mahlo considered even 'larger' cardinals, see for instance [77] and [80].

8.2 *Measurable cardinals*

From the side of abstract measure theory a condition on cardinals has proposed which turned out to require very large cardinals. A cardinal κ is called measurable if the set of all ordinals less than κ (i.e. κ) allows a measure m such that $m(X) \in \{0,1\}$ and $m(\{\alpha\}) = 0$, $m(\kappa) = 1$ and m is κ-additive.

Measurable cardinals were first studied by Ulam and Tarski [148]. It was shown by Ulam that every cardinal, less than the first strongly inaccessible cardinal, cannot be measurable and Tarski showed that the first inaccessible cardinal is also not measurable [145] (see also [146]).

It is an open problem whether the statement "There exists a measurable cardinal" is consistent with **ZF**.

The relation between the axiom of constructibility and the axiom "there is a measurable cardinal" has been determined by Scott in [128]; he showed that $V = L \vdash$ "there is no measurable cardinal" by applying ultra filter techniques.

Applications of measurable cardinals are found in topology and in algebra[78]).

Also measurable cardinals find applications in the model theory of infinitary languages (cf. [5]).

Azriel Levy presented in his Ph. D. Thesis (1958) an axiomschema (or rather: a collection of schemata) that turned out to have immediate consequences for the existence of larger cardinals[79]). The schemata are called *reflection principles* and they state that for each formula of **ZF** truth in the (or a) universe is equivalent to truth in a smaller domain. To be precise, let Scm (z) express that z is a standard complete (i.e. transitive) model satisfying the axioms of pairing, powerset and union. Then

$$\forall y \, \exists z [y \in z \wedge \mathrm{Smc}(z) \wedge \forall x_1 \in z \ldots \forall x_n \in z (A \leftrightarrow \mathrm{Rel}(z,A))]$$

Rel (z,A) is the formula obtained from A by restricting all quantifiers to the domain z. The variables $x_1, \ldots, x_n$ are those occurring free in A[80]).

---

[78]) See [146].

[79]) See [80], [9], [35].

[80]) This particular principle is called the *principle of complete reflection*, CR.

Levy showed that CR can replace in **ZF** the axioms of infinity and of replacement. Therefore this reflection principle does not add extra strength to the system **ZF**, by adding extra conditions the power of the principle can however be increased, for example the replacement of Scm (z) by "z is a standard model of **ZF**" the existence of 'large' cardinals can be shown.

9   Games and strategies

The axiom of choice is such a useful tool in the hands of mathematicians because it conjures up functions that would not be available otherwise. The applications mentioned in section 3 convincingly show the power of AC. There are however alternatives to AC, which may also have desirable existential consequences. Already in 1927 Church [27] considered alternatives to AC. A nice alternative with a plausible motivation was proposed by J. Mycielsky and H. Steinhaus in 1962 [93].

Their idea was based on game theory. Let us consider the following game between two persons I and II: A is a set of countable sequences of a set X. I and II alternatingly choose elements from X:

I   chooses x_0 x_2 x_4 x_6 etc.,
II   chooses x_1 x_3 x_5 x_7 etc.

I wins if the sequence $<x_i>$ belongs to A, otherwise II wins. Now a *strategy* is a mapping $f: \bigcup_{n \in \omega} X^n \to X$. f is a *winning strategy* for I if the strategy $<x_i>$, determined by

$$\begin{cases} x_0 = f(0) \\ x_{2n} = f(x_0, x_1, \ldots, x_{2n-1}), \text{ where } x_{2i+1} \text{ is arbitrarily chosen,} \end{cases}$$

belongs to A.

Likewise a winning strategy for II is defined. The game, determined by A and X is denoted by $\Gamma(X,A)$. $\Gamma(X,A)$ is called *determined* if there is a winning strategy for either I or II.

We can now formulate the *axiom of determinateness* (or *determinacy*) for X:

AD_X: Every game $\Gamma(X,A)$ is determined.

Mycielski noted that the axioms of determinateness for $X = \{0,1\} = 2$ and $X = \omega$ are equivalent [94].

Some consequences of this new axiom are given by:

(i) $\mathbf{ZF} + AD_2 \vdash AC^\omega$ (countable choice).
(ii) $\mathbf{ZF} + AD_2 \vdash$ 'Every set of reals is Lebesgue-measurable'.

Taking into account Vitali's result (p. 28) that $\mathbf{ZF} + AC \vdash$ 'There is a set of reals which is not Lebesgue-measurable', we conclude that $\mathbf{ZF} + AD_2 \vdash \neg AC$.

However the axiom of countable choice is still available, if one accepts the axiom of determinateness, so enough of an analysis can be handled to make everyday mathematics possible.

(iii) $\mathbf{ZF} + AD_2 \vdash$ 'Every non-denumerable set of reals contains a perfect subset'.

Under assumption of the axiom of determinateness some cardinals have quite unusual properties, Specker, for instance, showed that $\aleph_1$ is inaccessible. Solovay even showed that $\mathbf{ZF} + AC +$ 'there exists a measurable cardinal' is consistent if $\mathbf{ZF} + AD_2$ is consistent.

Of course one can ask for which sets A $\Gamma(X,A)$ is determined, given a fixed X. In the case of ω^ω (Baire space) Morton Davis showed that the game $\Gamma(X,A)$ is determined for every A that is $G_{\delta\sigma}$ or $F_{\sigma\delta}$ and Paris extended the results to sets A in $G_{\delta\sigma\delta}$ [81]). The axiom of determinateness proves to be very useful in the context of hierarchy theory.

---

[81] We say that a set is F_σ (G_σ) if it is the countable union (intersection) of closed (open) sets. A set is $G_{\delta\sigma}$ ($F_{\sigma\delta}$) if it is the countable union (intersection) of sets in G_δ (F_σ).
A set is $G_{\delta\sigma\delta}$ if it is the countable intersection of sets in $G_{\delta\sigma}$.

Bibliography*)

[1] ACKERMANN, W. Die Widerspruchsfreiheit der allgemeinen Mengenlehre. *Math. Ann. 114* (1937) pp. 305-315.

[2] ADDISON, J. W. Some consequences of the axiom of constructibility. *Fund. Math. 46* (1959) pp. 337-357.

[3] BALLAUF, L. Review of G. Cantor, Grundlagen einer allgemeinen Mannigfaltig-keitslehre. *Zeitschrift für exakte Philosophie 12* (1883) pp. 375-395.

[4] BELL, E. *Men of mathematics, II* London (1937).

[5] BELL, J. L. and A. B. SLOMSON. *Models and ultraproducts.* Amsterdam (1969).

[6] BENACERRAF, P. and H. PUTNAM (ed.). *Philosophy of mathematics*, selected readings. Englewood Cliffs (1964).

[7] BERNAYS, P. A system of axiomatic set theory. *J.S.L. 2* (1937) pp. 65-77, *J.S.L. 6* (1941) pp. 1-17, *J.S.L. 7* (1942) pp. 65-89, *J.S.L. 8* (1943) pp. 89-106, *J.S.L. 13* (1948) pp. 65-79, *J.S.L. 19* (1954) pp. 81-96.

[8] BERNAYS, P. and A. A. FRAENKEL. *Axiomatic set theory.* Amsterdam (1958).

[9] BERNAYS, P. Unendlichkeitsschemata in der axiomatischen Mengenlehre. In *Essays on the foundations of mathematics.* Jerusalem (1961).

[10] BERNSTEIN, F. Über die Reihe der transfiniten Ordnungszahlen. *Math. Ann. 60* (1905) pp. 187-193.

[11] BERNSTEIN, F. Zum Kontinuumproblem. *Math. Ann. 60* (1905) pp. 463-464.

[12] BETH, E. W. Semantic construction of intuitionistic logic. *Mededelingen der Kon. Ned. Ak. v. Wet. 18* (1956) no. 11.

[13] BETH, E. W. *The foundations of mathematics.* Amsterdam (1959).

*)   J. S. L. stands for Journal of Symbolic Logic.

[14] BOCKSTAELE, P. *Het intuïtionisme bij de Franse wiskundigen.* Brussel (1949).

[15] BOFFA, M. Graphes extensionnels et axiome d'universalité. *Zeitschr. für Math. Logik und Grundl. der Math. 14* (1968) pp. 329-334.

[16] BOFFA, M. Les ensembles extraordinaires. *Bull. Soc. Math. Belgique 20* (1968) pp. 3-15.

[17] BOIS-REYMOND, P. DU. Über asymptotische Werte, infinitäre Approximationen und infinitäre Auflösung von Gleichungen. *Math. Ann. 8* (1875) pp. 363-414.

[18] BOLZANO, B. *Paradoxien des Unendlichen.* Leipzig (1851), Darmstadt (1964).

[19] BOREL, E. *Leçons sur la théorie des fonctions.* Paris (1898).

[20] BOREL, E. Quelques remarques sur les principes de la théorie des ensembles. *Math. Ann. 60* (1905) pp. 194-195.

[21] BROUWER, L. E. J. *Over de grondslagen der wiskunde.* (On the foundations of mathematics). Amsterdam (1907). An English translation is to appear in the collected works.

[22] BROUWER, L. E. J. Beweis der Invarianz der Dimensionszahl. *Matth. Ann. 70* (1910) pp. 305-313.

[23] BROUWER, L. J. E. Zur Begründung der intuitionistischen Mathematik I. *Math. Ann. 93* (1925) pp. 244-257.

[24] BROUWER, L. J. E. Zur Begründung der intuitionistischen Mathematik III. *Math. Ann. 96* (1926) pp. 451-489.

[25] CANTOR, G. *Gesammelte Abhandlungen,* edited by E. Zermelo. Berlin (1932). Reprinted Hildesheim (1966).

[26] CANTOR, G. Über ein neues und allgemeines Kondensationsprinzip der Singularitäten von Funktionen. *Math. Ann. 19* (1882) pp. 588-594, also in [25].

[27] CHURCH, A. Alternatives to Zermelo's assumption. *Trans Am. Math. Soc. 29* (1927) pp. 178-208.

[28] COHEN, P. J. A minimal model for set theory. *Bull .Am. Math. Soc. 69* (1963) pp. 537-540.

[29] COHEN, P. J. The independence of the continuum hypothesis I, II. *Proc. Nat. Ac. Sci. USA 50* (1963) pp. 1143-1148, *51* (1964) pp. 105-110.

[30] COHEN, P. J. *Set theory and the continuum hypothesis.* New York (1966).

[31] DALEN, D. VAN. A note on spread cardinals. *Compositio Mathematica 20* (1968) pp. 21-28.

[32] DAVIS, M. *The undecidable.* N.Y. (1965).

[33] DEDEKIND, R. *Was sind und sollen die Zahlen?* Brunswick (1888). Also in [34].

[34] DEDEKIND, R. *Gesammelte Mathematische Werke III.* Braunschweig (1932).

[35] FELGNER, U. *Models of ZF-set theory.* Springer lecture notes (1971).

[36] FRAENKEL, A. A. Der Begriff 'definit' und die Unabhängigkeit des Auswahlaxioms. *Sitzungsberichte der Preussische Akademie der Wissenschaften, physikalische Klasse* (1922) pp. 253-257. Also in Van Heyenoort [61] pp. 284-289.

[37] FRAENKEL, A. A. Zu den Grundlagen der Cantor-Zermeloschen Mengenlehre. *Math. Ann. 86* (1922) pp. 230-237.

[38] FRAENKEL, A. A. *Einleitung in die Mengenlehre.* Berlin (1928).

[39] FRAENKEL, A. A. and Y. BAR-HILLEL. *Foundations of set theory.* Amsterdam (1958).

[40] FREGE, G. *Begriffsschrift,* eine der arithmetischen nachgebildete Formelsprache des reinen Denkens. Halle (1879).

[41] FREGE, G. *Die Grundlagen der Arithmetik,* eine logisch-mathematische Untersuchung über den Begriff der Zahl. Breslau (1884), Hildesheim (1961). English translation by J. L. Austin. Harper Torchbooks (1960).

[42] FREGE, G. *Die Grundgesetze der Arithmetik.* Jena (1893, 1903), Hildesheim (1962).

[43] FREGE, G. *Begriffsschrift und andere Aufsätze,* ed. I. Angelelli. Hildesheim (1964).

[44] GALILEI, G. *Discorsi e Dimostrazioni Matematiche intorno a due nuove scienze.* Leiden (1638). English translation in Dover Publications, N.Y. German translation in Ostwald's Klassiker Nr. 11.

[45] GAUSS, C. F. *Gesammelte Abhandlungen,* Bd. VIII. Berlin (1932).

[46] GENTZEN, G. Beweisbarkeit und Unbeweisbarkeit von Anfangsfällen der transfiniten Induktion in der reinen Zahlentheorie. *Math. Ann. 119* (1943) pp. 140-161.

[47] GÖDEL, K. The consistency of the axiom of choice and the generalized continuum hypothesis. *Proc. Nat. Ac. Science 24* (1938) pp. 556-557.

[48] GÖDEL, K. Consistency proof for the generalized continuum hypothesis. *Proc. Nat. Ac. Science 25* (1939) pp. 220-224.

[49] GÖDEL, K. *The consistency of the continuum hypothesis.* Princeton (1940).

[50] GÖDEL, K. What is Cantor's continuum problem? *Am. Math. Monthly 54* (1947) pp. 515-525. Also in [6] (revised and expanded).

[51] GRATTAN-GUINESS, I. An unpublished paper by Georg Cantor: Prinzipien einer Theorie der Ordnungstypen. Erste Mitteilung. *Acta Mathematica 124* (1970) pp. 65-107.

[52] GRZEGORCZYK, A. A plausible formal interpretation of intuitionistic logic. *Indag. Math. 26* (1964) pp. 596-601.

[53] HÁJEK, P. Modelle der Mengenlehre in denen Mengen gegebener gestalt existieren. *Zeitschr. für Math.-Logik und Grundl. der Math. 11* (1965) pp. 103-115.

[54] HAMEL, G. Ein Basis aller Zahlen und die unstetige Lösungen der Funktionalgleichung $f(x+y)=f(x)+f(y)$. *Math. Ann. 60* (1905) pp. 459-462.

[55] HARTOGS, F. Über das Problem der Wohlordnung. *Math. Ann. 76* (1915) pp. 438-443.

[56] HAUSDORFF, F. Der Potenzbegriff in der Mengenlehre. *Jahresberichte der DMV 13* (1904) pp. 570 ff.

[57] HAUSDORFF, F. Grundzüge einer Theorie der geordnete Mengen. *Math. Ann. 64* (1908) pp. 443-505.

[58] HAUSDORFF, F. *Nachgelassene Schriften* II. Stuttgart (1969).

[59] HELMHOLTZ, H. Zählen und Messen. In *Philosophische Aufsätze*, E. Zeller zu 50-jährigen Doktorjubiläum gewidmet. Leipzig (1887) p. 76.

[60] HESSENBERG, G. *Grundbegriffe der Mengenlehre*. Abhandlungen der Fries'schen Schule (1906).

[61] HEYENOORT, J. VAN. *From Frege to Gödel*, a source-book in mathematical Logic 1879-1931. Cambridge, Mass. (1967).

[62] HILBERT, D. Mathematische Probleme. Vortrag, gehalten auf dem internationalen Mathematiker-Kongress zu Paris 1900. *Archiv der Math. und Physik, 3rd series 1* (1901) pp. 44-63, 213-237.
Also in [65] p. 290 ff.

[63] HILBERT, D. Über das Unendliche. *Math. Ann. 95* (1926) pp. 161-190. Also in Van Heyenoort [61] pp. 367-392.

[64] HILBERT, D. Die Grundlagen der Mathematik. *Abh. a.d. Math. Sem. d. Hamb. Universität 6* (1928) pp. 65-85. Also in Van Heyenoort [61] pp. 464-479.

[65] HILBERT, D. *Gesammelte Abhandlungen* Vol. 3. Berlin (1935).

[66] JENSEN, R. *Modelle der Mengenlehre*. Berlin (1967).

[67] KELLEY, J. L. *General topology*. Princeton (1955).

[68] KLEENE, S. C. *Introduction to meta-mathematics*. Amsterdam-New York (1952).

[69] KLIMOVSKY, G. The axiom of choice and the existence of maximal abelian subgroups. *Rev. Un. Math. Argentina 20* (1962) pp. 267-287.

[70] KNEEBONE, G. T. *Mathematical logic and the foundations of mathematics*. London (1963).

[71] KÖNIG, J. Zum Kontinuumproblem. *Math. Ann. 60* (1905) pp. 177-180. Berichtigung ibid p. 462.

[72] KÖNIG, J. Über die Grundlagen der Mengenlehre und das Kontinuumproblem. *Math. Ann. 61* (1906) pp. 156-160. Also in Van Heyenoort [61] pp. 145-149.

[73] KOWALEWSKI, G. *Bestand und Wandel*. München (1950).

[74] KRIPKE, S. Semantical Analysis of intuitionistic logic I. In *Formal systems and recursive functions*. Amsterdam (1965) pp. 92-130.

[75] KRONECKER, L. Über den Zahlbegriff. *Journal für die reine und angewandte Mathematik 101* (1887) pp. 337-355.

[76] KRONECKER, L. *Jahresbericht der Deutsche Mathematiker-Vereinigung 1* (1890) pp. 23-25.

[77] KURATOWSKI, K. and A. MOSTOWSKI. *Set theory*. (1968). Warszawa-Amsterdam

[78] LEBESGUE, H. Contribution a l'étude des correspondances de M. Zermelo. *Bull. de la Soc. Math. de France 35* (1907) pp. 202-212.

[79] LEIBNIZ, G. *Mathematische Schriften* III, ed. Gerhardt (1855, 1856) pp. 536, 563.

[80] LÉVY, A. Axiom schemata of strong infinity in axiomatic set theory. *Pac. J. Math. 10* (1960) pp. 223-238.

[81] Lévy, P. Remarques sur un théorème de Paul Cohen. *Revue de Métaphysique et de Morale 69* (1964) pp. 88-95.

[82] Mahlo, P. Über lineaire transfinite Mengen. *Leipziger Berichte math. phys. Klasse 63* (1911) pp. 187-225.

[83] Mendelson, E. The axiom of Fundierung and the axiom of choice. *Arch. Math. Logik und Grundlagenforschung 4* (1958) pp. 65-70.

[84] Mendelson, E. *Introduction to mathematical logic* (especially Ch. 4) Princeton (1964).

[85] Meschkowski, H. *Probleme des Unendlichen.* Werk und Leben Georg Cantors. Braunschweig (1967).

[86] Mirimanoff, D. Les antinomies de Russell et de Burali-Forti et le problème fondemental de la théorie des ensembles. *L'enseignement mathématique 19* (1917) pp. 37-52.

[87] Monk, D. *Introduction to set theory.* N.Y. (1969).

[88] Montague, R. Semantical closure and non-finite axiomatizability I. In *Infinitistic Methods.* Warszawa (1961).

[89] Morse, A. *A theory of sets.* N.Y. (1965).

[90] Mostowski, A. Über die Unabhängigkeit des Wohlordnungssatzes vom Ordnungs-prinzip. *Fund. Math. 32* (1939) pp. 201-252.

[91] Mostowski, A. *Thirty years of foundational studies.* Oxford (1966).

[92] Mostowski, A. *Constructible sets with applications.* Warszawa-Amsterdam (1969)

[93] Mycielski, J. and H. Steinhaus. A mathematical axiom contradicting the axiom of choice. *Bull. Ac. Polon. Sci. Ser. Sci. Math. Astr. Phys. 10* (1962) pp. 1-3.

[94] Mycielski, J. On the axiom of determinateness. *Fund. Math. 53* (1964) pp. 205-224.

[95] Neumann, J. von. Zur Einführung der transfiniten Zahlen. *Acta literarum ac scientiarum Regiae Universitatis Hungariae Francisco Josephina. Sectio scientiarum math. 1* (1923) pp. 199-208. Also in Van Heyenoort [61] pp. 346-354.

[96] Neumann, J. von. Eine Axiomatisierung der Mengenlehre. *Journal für die reine und angewandte Math. 154* (1925) pp. 219-240. Also in Van Heyenoort [61] pp. 393-413.

[97] Neumann, J. von. Eine Axiomatisierung der Mengenlehre. *Math. Zeitschrift 27* (1928) pp. 669-752.

[98] Neumann, J. von. Über eine Widerspruchsfreiheitsfrage in der axiomatischen Mengenlehre. *Journ. für die reine und angewandte Math. 160* (1929) pp. 227-241. Also in [99] pp. 494-508

[99] Neumann, J. von. *Collected works.* Vol. I. Oxford (1961).

[100] Noether, E., ed. J. Cavaillès. *Briefwechsel Cantor-Dedekind,* Paris (1937). Also in J. Cavaillès. *Philosophie mathématique.* Paris (1962).

[101] PEANO, G. Démonstration de l'intégrabilité des équations différentielles ordinaires. *Math. Ann. 37* (1890) pp. 182-288.

[102] PEANO, G. *Formulaire de mathématiques.* Vol. 1, 2, 3, 4. Turin (1895-1903).

[103] PEANO, G. Additione. *Revista del Mathematica 8* (1906) pp. 143-157.

[104] PEANO, G. Sulla definizione di funzione. *Atté della Reale Accademia dei lincei. Classe di scienze fisiche, mathematische e naturali 20* (1911) pp. 3-5.

[105] PIERCE, C. S. On the algebra of logic, a contribution to the philosophy of notation. *Am. J. Math. 7* /1885) pp. 180-202.

[106] PINL, M. Kollegen in einer dunklen Zeit. *Jahresberichte Deutsche Mathematiker Vereinigung 71* (1969) pp. 167-228.

[107] POINCARÉ, H. *Science et méthode.* Paris (1908). English translation in Dover Publications.

[108] POINCARÉ, H. *Sechs Vorträge über ausgewählte Gegenstände aus der reinen Mathematik und mathematischen Physik.* Leipzig (1910)

[109] PROCEEDINGS *of the AMS.* Summer institute on axiomatic set theory (1967) UCLA (1971).

[110] PROCEEDINGS *of the International Congress of Mathematicians,* ed. I. G. Petrovsky. Moscow (1966).

[111] QUINE, W. V. O. *Set theory and its logic.* Cambridge, Mass (1963).

[112] RICHARD, J. Les principes des mathématiques et le problème des ensembles. *Revue générale des sciences pures et appliquées 16* (1905) pp. 541. Also in van Heyenoort [61] pp. 142-144.

[113] RIEGER, L. *Chechoslovak Math. J. 7* (1957) pp. 323-357.

[114] ROBINSON, R. M. The theory of classes, a modification of Von Neumann's system. *J.S.L. 2* (1937) pp. 29-32.

[115] ROGERS, H. JR. *Theory of recursive functions and effective computability.* New York (1967).

[116] ROSSER, J. B. *Simplified independence Proofs.* Boolean valued models of set theory. N.Y. (1969).

[117] RUBIN H. and J. E. RUBIN. *Equivalents of the axiom of choice.* Amsterdam (1963).

[118] RUBIN, J. E. *Set theory for mathematicians.* Amsterdam (1967).

[119] RUSSELL, B. On some difficulties in the theory of transfinite numbers and order types. *Proc. of the London Math. Soc. 4* (1907) pp. 29-53.

[120] SCHILP, ed. *The philosophy of Bertrand Russell.* New York (1944).

[121] SCHOENFLIESS, A. *Entwicklung der Mengenlehre und ihrer Anwendungen.* Leipzig und Berlin, (1900) 2nd ed. (1913).

[122] SCHOENFLIESS, A. Die Krisis in Cantor's Mathematischem Schaffen. *Acta Mathematica 50* (1927) pp. 1-23.

[123] SCHOENFLIESS, A. Zur Erinnerung an Georg Cantor. *Jahresberichte DMV 31* (1922) pp. 97-106.

[124] SCHRÖDER, E. *Nova Acta Leopoldina 71* (1898).

[125] SCHÜTTE, K. *Beweistheorie*. Berlin (1960).

[126] SCOTT, D. Prime ideal theorems for rings, lattices and Boolean algebras. *Bull. Am. Soc. 60* (1954) pp. 390.

[127] SCOTT, D. *The notion of rank in set-theory*. Summaries of Talks – Summer Institute for Symbolic Logic, Cornell University (1957).

[128] SCOTT, D. Measurable cardinals and constructible sets. *Bull. Ac. Polon. Sci. Sér. Sci. Math. Astr. Phys. 9* (1961) pp. 521-524.

[129] SCOTT, D. A proof of the independence of the continuum hypothesis. *Math. Systems Theory 1* (1967) pp. 89-111.

[130] SHEPERDSON, J. Inner models for set theory I, II, III. *J.S.L. 16* (1951) pp. 161-190, *J.S.L. 17* (1952) pp. 225-237. *J.S.L. 18* (1953) pp. 145-167.

[131] SHOENFIELD, J. R. *Mathematical logic*. Reading, Mass (1967).

[132] SIERPINSKI, W. Sur une hypothèse de M. Lusin. *Fund. Math. 25* (1935) pp. 132-135.

[133] SIERPINSKI, W. L'hypothèse géneralisé du continu et l'axiome du choix. *Fund. Math. 34* (1947) pp. 1-5.

[134] SKOLEM, Th. *Einige Bemerkungen zur axiomatischen Begründung der Mengenlehre*. Matematikerkongressen i Helsingfors den 4-7 Juli 1922. Den femte Skandinawiska matematiker-kongressen, Redogörelse. Also in [136] pp. 137-152 and van Heyenoort [61] pp. 290-301.

[135] SKOLEM, TH. Einige Bemerkungen zu der Abhandlung von E. Zermelo: "Über die Definitheit in der Axiomatik". *Fund. Math. 15* (1930) pp. 337-341. Also in [136] pp. 275-279.

[136] SKOLEM, TH. Über die Nichtcharakterisierbarkeit der Zahlenreihe mittels endlich oder abzählbar unendlich vieler Aussagen mit ausschliesslich Zahlenvariablen. *Fund. Math. 23* (1934) pp. 150-161. Also in [136] pp. 355-366.

[137] SKOLEM, TH. *Selected works in logic*, ed. J. E. Fenstad. Oslo (1970).

[138] SOLOVAY, R. $2^{\aleph_0}$ can be anything it ought to be. In *The theory of models*. Amsterdam (1965) pp. 433-434.

[139] SOUSLIN, N. Problème 3. *Fund. Math. 1* (1920) p. 223.

[140] SPECKER, E. Verallgemeinerte Kontinuumshypothese und Auswahlaxiom. *Archiv der Math. 5* (1954) pp. 332-337.

[141] SPECKER, E. Zur Axiomatik der Mengenlehre (Fundierungs- und Auswahlaxiom). *Zeitschrift für math. Logik u. Grundlagen d. Math. 3* (1957) pp. 173-210.

[142] STEINITZ, E. *Theorie der algebraischen Körper*, Berlin, 1930.

[143] TARSKI, A. and A. LINDENBAUM. Communication sur les recherches de la théorie des ensembles. *Comptes rendus Soc. Sci. et Lettres de Varsovie, Classe III 19* (1926) pp. 299-330.

[144] TARSKI, A. Über unerreichbare Kardinalzahlen. *Fund. Math. 30* (1938) pp. 68-89.

[145] TARSKI, A. Some problems and results relevant to the foundation of set theory. *Proc. 1960 Intern. Congr. Logic Math. and Phil. Science.* Stanford (1962) pp. 125-135.

[146] TARSKI, A. and J. H. KEISLER. From accessible to inaccessible cardinals. *Fund. Math. 53* (1964) pp. 117-199.

[147] TROELSTRA, A. S. *Principles of intuitionism.* Berlin (1969).

[148] ULAM, S. Zur Masstheorie in der allgemeinen Mengenlehre. *Fund. Math. 15* (1930) pp. 140-150.

[149] VITALI, G. *Sul problema della misura dei gruppi di punti di una retta.* Bologna (1905).

[150] WANG, HAO. *A survey of mathematical logic.* Peking-Amsterdam (1963).

[151] WITTGENSTEIN, L. *Bemerkungen über die Grundlagen der Mathematik.* Oxford (1956) (with English translation).

[152] YOUNG, W. H. and G. C. YOUNG. Review of "The theory of functions of a real variable and the theory of Fourier series" by E. W. Hobson. *Math. Gazette 14* (1928) pp. 98-104.

[153] ZERMELO, E. Beweis das jede Menge wohlgeordnet werden kann. *Math. Ann. 59* (1904) pp. 514-516. Also in van Heyenoort [61] p. 139.

[154] ZERMELO, E. Neuer Beweis für die Möglichkeit einer Wohlordnung. *Math. Ann. 65* (1908) pp. 107-128. Also in [61] pp. 183-198.

[155] ZERMELO, E. Untersuchungen über die Grundlagen der Mengenlehre I, *Math. Ann. 65* (1908) pp. 261-281. Also in van Heyenoort [61] pp. 199-215.

[156] ZERMELO, E. Über den Begriff der Definitheit in der Axiomatik. *Fund. Math. 19* (1929) pp. 339-344.

[157] ZERMELO, E. Über Grenzzahlen und Mengenbereiche. *Fund. Math. 16* (1930) pp. 29-47.

The integral from Riemann to Bourbaki

"...l'interêt mathématique de ces études (historiques) est de permettre de reconnaître l'étroite parenté qui unit des recherches effectuées à plus de vingt siècles de distances."

HENRI LEBESGUE [38] p. 11.

1 Introduction; the period before Riemann

Although I have the intention to write about the development of the integral in the period from Riemann into modern times, it would not be justified to pass over the two thousand years before Riemann in which several famous mathematicians studied this notion. By omitting entirely this long period, there would be no opportunity to mention the names of Euclides, Eudoxos, Archimedes, and, centuries later, Leibniz, Newton; without pretending that this list should be complete. This is all the more unjustified as there are several lines, leading from Antiquity to modern times. Therefore, a short introduction

HENRI LEBESGUE

is necessary; I shall mention some ideas which return in our modern theories. In this introduction completeness is not aimed at.

1.1 *Greek mathematics*

What we now call the integral calculus was, in Greek Antiquity, the problem of calculating volumes and areas, which, for the rest, remained so for centuries. Nowadays we speak about measure-theory and set-functions. The Greek mathematicians succeeded in several cases to calculate volumes and areas. I will mention some of their ideas and methods.

First I must mention EUCLIDES, who probably lived in a period about 300 b.C. (this is not known exactly). In his famous book "The Elements", he gives a method for calculating several volumes, for instance for prisms and pyramids, but he already considers the circle too.

Perhaps more important for our purpose as a precursor for integral calculus is ARCHIMEDES (287-212). He wrote about the sphere and the cylinder, the quadrature of the parabola and in determining volumes and areas and barycentres with methods, in a certain sense comparable to integration-theory, he attained exactness.

Among the methods used in Antiquity I must first mention the so called *Exhaustion method*. I will give some more details about this method, because the principle on which this method is founded returns in other parts of mathematics. Let us give an example.

Let I be the volume of a tetrahedron and V the volume of a prism with equal height and base. One shows that $I = \frac{1}{3}V$ by showing that $I > \frac{1}{3}V$ and $I < \frac{1}{3}V$ are both contradictory and for this the exhaustion method is used. The scheme of a proof according to this method is about as follows (there are modifications). Suppose it is to be proved that $A = B$.

Proof. 1.   Suppose $A < B$. Then there is $\varepsilon > 0$ such that $B - A = \varepsilon$. However there is an A_r such that

$$A_n < A,$$
$$|B - A_n| < \varepsilon.$$

It follows $A > A_n > B - \varepsilon = A$ and this is a contradiction.

2.   Suppose $A > B$; this leads in an analogous way to a contradiction.

3. $A = B$ follows now from 1 and 2.

The construction of A_n is performed with the exhaustion method[1]). For a tetrahedron

---

[1])   The name exhaustion method appears first in the work of Grégoire de Saint Vincent (1647). Dijksterhuis [39] p. 68, remarks that this name is fundamentally wrong. Precisely because it is a limiting

for instance this is done by constructing prisms inscribed in the tetrahedron and prisms circumscribing the tetrahedron. In this way the quantity to be determined is obtained by approximating it from above and from below; it is in principle a limiting process.

The exhaustion method is related to the theory of ratio and proportion, which goes back to EUDOXOS (408-355).

The way of proving results with this method is based on an axiom, which is usually named after Archimedes, but which could better be called the axiom of Eudoxos. In modern mathematics one gives the following *definition*:

Let G be a partially ordered group (in multiplicative notation). The order is said to be archimedean if

$$a^n < b \ (n = 0, \pm 1, \pm 2, \ldots), \ a,b \in C, \ implies \ a = e.$$

This is a definition, that is to say, some ordered groups may be archimedean, others not. The axiom of Eudoxos says that the set of real numbers is archimedean in its natural ordering. This notion plays an important role in modern mathematics, for instance in the theory of lattices; orderings which are not archimedean are also used.

The method of approximation from above and from below occurs in several places in modern mathematics. I mention for instance the introduction of irrational numbers by means of Dedekind-cuts (where it is a means of definition), in integration theory the Riemann upper and lower integral and the theory of the Perron-integral (which I shall treat later on in some detail). Even in the theory of differential equations the method is used: I mention the method of Perron in the theory of the problem of Dirichlet, by which certain harmonic functions (i.e. solutions of $\Delta u = 0$) are defined with the aid of subharmonic and superharmonic functions. This method runs through mathematics as a thread from Antiquity to modern times.

It must be mentioned, however, that the exhaustion method provided the Greek mathematicians with a means of proving in an exact way results which were already known in some way or other. It seems that methods of a more or less empirical character were used to obtain results. An example is the mechanical method which was used by Archimedes. In this he considers an area as being composed of line segments, which in an appropriate way are used to construct a lever in a state of equilibrium. The equilibrium conditions led him to the calculation of an area. This method must have led him in the dangerous domain of the geometrical atoms, the "indivisibile", and thus in the attitude of the Greek mathematicians towards infinity. A surface or a volume was considered to be composed of geometrical atoms; one had to determine the "sum" of these atoms. This

---

process, the area of a sphere, for example, cannot be exhausted by inscribed polyhedrons with an increasing number of sides. The name compression method is proposed by Dijksterhuis in his work on Archimedes [40], p. 130.

79

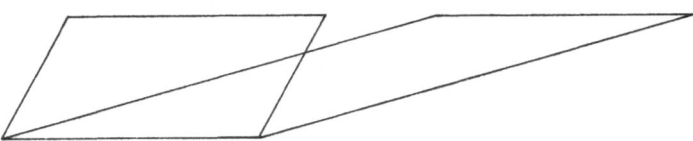

is a domain which is related to Zeno's paradox concerning Achilles and the tortoise (which has nothing to do with the paradoxes of the preceding chapter). Zeno's paradoxes showed that the hypothesis that the number of atoms was finite as well as the hypothesis that this number was infinite led to difficulties. This is perhaps the reason that Archimedes did not trust his mechanical method and so the Greek came to formulate exact proofs by means of the exhaustion method.

In our mathematics the solution of Zeno's paradox is of course expressed by the fact that a segment can be obtained as the sum of an infinite number of segments each with a finite length.

It is not my intention to go deeper into Greek mathematics and the philosophical considerations which are related to it (for more details see for instance Struik [103]).

I only want to make a final remark concerning the geometrical atoms. Even in modern times the method of atoms is used. For instance in differential geometry, where in first approximation a small part of a surface is considered as being flat. Or in physics when one is concerned with the construction of the differential equations for a certain phenomenon. There are also the methods of discretization for the solution of problems.

There is an other method, used in Antiquity, which I have to mention because of its relations to modern integration theory, although it is perhaps more in the domain of the foundations of geometry. It concerns the problem of the introduction of the concepts of area and volume in elementary mathematics, especially for polygons and polyhedrons. This method, used by the Greek mathematicians, is that of defining area and volume by means of the juxtaposition (addition) of elementary geometrical figures and the application of deplacements in order to obtain results for more complicated figures. It is a means for obtaining results like: the figures A and B are equal in the sense that their ratio equals 1. It must be well distinguished from the mapping of geometrical figures into the set of positive real numbers. A proposition "A and B have equal area" means here: "A and B are composed of the same number of pairwise congruent figures". I call this method the "*decomposition method*". In modern terms, it is the problem of defining an equivalence relation for polygons and polyhedrons. This method is still used in school mathematics for proving, for instance, that parallelograms with equal base and height have equal area (without associating a positive real number to such figures). It is to be kept in mind that geometrical figures with equal area are not necessarily congruent.

Euclides gives several propositions of this kind; compare for instance the theorem of Pythagoras, expressed in terms of areas of triangles. He based his theory on a number of axioms, among which I mention Axiom VII "Things which can cover each other are equal". They are so adjusted that an equivalence relation is obtained. In modern terms, such a theory must be formulated with transformation groups. I give some definitions in $\mathbf{R}^3$ (they are the same for polygons in $\mathbf{R}^2$).

Let G be a transformation group in $\mathbf{R}^3$, containing the translations. The polyhedrons A and B are said to be *G-equivalent*, if they can be decomposed in a finite number of polyhedrons which are pairwise equal with respect to G, that is can be transformed pairwise in each other by means of elements of G.

The polyhedrons A and B are said to be *weak G-equivalent* if there are G-equivalent polyhedrons C and D such that $A + C$ and $B + D$ are G-equivalent[2]).

The classical case is obtained when one takes for G the full transformation group.

Evidently this defines an equivalence relation in the set of polyhedrons. It is clear that this point of view is quite different from the exhaustion method. But there is a thread from it into modern mathematics: we find it again in the problem of the existence of a measure (set function), defined on an appropriate system of sets, which is invariant under the application of the transformations of a transformation group. This is the problem of *Haar-measure* on locally compact groups; I will return to it later on.

It is beyond the purpose of this book to treat in detail the problems which derive from these definitions, but there is, however, an interesting question which I shall mention.

It is trivial that equivalent polyhedrons are weak-equivalent. That, conversely, weak-equivalent polyhedrons are equivalent was only proved in recent times[3]). (See Sydler [104] Hadwiger [54]).

It is clear that equivalent polyhedrons (in $\mathbf{R}^2$ polygons) have the same volume (this last notion is defined as in elementary mathematics). But what about the converse? There is a remarkable difference in this respect between the spaces $\mathbf{R}^2$ and $\mathbf{R}^3$. In $\mathbf{R}^2$ the converse is true: two polygons with equal area are equivalent. Hilbert treated this problem in his "Grundlagen der Geometrie" [58]. He proved that the theory of area in the plane can be based axiomatically on the concept of weak-equivalence (Ergänzungsgleichheit) of polygons[4]).

---

[2]) What we call equivalent is often called "zerlegungsgleich" and what we call weak-equivalent is "ergänzungsgleich".

[3]) It is for historical reasons that I have introduced here the notion of weak-equivalent polyhedrons. For an excellent survey of the development of area and volume in elementary geometry I refer the reader to the article "Über die Lehre von der Äquivalenz (Gleichheit)" in the book of Enriques [41].

[4]) Such a foundation is possible without using the axiom of Archimedes-Eudoxos. In nor-archmedean geometry, however, there exist triangles, as Hilbert showed, which are weak-equivalent but not equivalent. See [58] for information on the role of the axiom of Archimedes-Eudoxos.

However, the converse is not true in $\mathbf{R}^3$, that is to say there exist polyhedrons with the same volume which are not equivalent. This has also been proved but only in recent times. Knowing this, it is understandable for us that the Greek mathematicians did not treat problems of volume of pyramids etc. by means of equivalence. But I don't know whether they have ever tried to do this.

Even Gauss has studied this problem, but he could not solve it. He writes about the problem in two letters to Gerling (an astronomer, pupil and friend of Gauss) dated 8 April 1844 and 17 April 1844[5]). In 1900 Hilbert attacks the problem again at the International Congres of mathematicians, held in Paris under the chairmanship of Henri Poincaré[6]). At this congres Hilbert presented a lecture, in which he proposes a number of problems to the mathematical world. One of these 23 problems concerns the problem about polyhedrons, mentioned above. Hilbert began his lecture as follows[7]).

" **17. Mathematische Probleme.**

[Archiv f. Math. u. Phys. 3. Reihe, Bd. 1, S. 44-63; S. 213-237 (1901)]

Wer von uns würde nicht gern den Schleier lüften, unter dem die Zukunft verborgen liegt, um einen Blick zu werfen auf die bevorstehenden Fortschritte unserer Wissenschaft und in die Geheimnisse ihrer Entwicklung während der künftigen Jahrhunderte! Welche besonderen Ziele werden es sein, denen die führenden mathematischen Geister der kommenden Geschlechter nachstreben? Welche neuen Methoden und neuen Tatsachen werden die neuen Jahrhunderte entdecken — auf dem weiten und reichen Felde mathematischen Denkens?

Die Geschichte lehrt die Stetigkeit der Entwicklung der Wissenschaft. Wir wissen, daß jedes Zeitalter eigene Probleme hat, die das kommende Zeitalter löst oder als unfruchtbar zur Seite schiebt und durch neue Probleme ersetzt. Wollen wir eine Vorstellung gewinnen von der mutmaßlichen Entwicklung mathematischen Wissens in der nächsten Zukunft, so müssen wir die offenen Fragen vor unserem Geiste passieren lassen und die Probleme überschauen, welche die gegenwärtigen Wissenschaft stellt, und deren Lösung wir von der Zukunft erwarten. Zu einer solchen Musterung der Probleme scheint mir der heutige Tag, der an der Jahrhundertwende liegt, wohl ge-

---

[5]) Carl Friedrich Gauss, Werke B8, p. 241, 244.

[6]) This was the second International Congres. The first was held in 1897 in Zürich.

[7]) David Hilbert, Gesammelte Abhandlungen, Bd. III, p. 290. I remark that not all these problems are solved untill yet (for instance problems about prime numbers and the zeros of the Z-function of Riemann).

eignet; denn die großen Zeitabschnitte fordern uns nicht bloß auf zu Rück-
blicken in die Vergangenheit, sondern sie lenken unsere Gedanken auch auf
das unbekannte Bevorstehende."

It is the third problem which concerns the polyhedrons and the foundations of mathematics. Hilbert writes about it in the following way:

"Aus dem Gebiete der Grundlagen der Geometrie möchte ich zunächst das
folgende Problem nennen.

3. Die Volumengleichheit zweier Tetraeder von gleicher Grundfläche und Höhe

Gauss spricht in zwei Briefen an Gerling sein Bedauern darüber aus,
daß gewisse Sätze der Stereometrie von der Exhaustionsmethode, d.h. in der
modernen Ausdrucksweise von dem Stetigkeitsaxiom (oder von dem Archime-
dischen Axiome) abhängig sind. Gauss nennt besonders den Satz von Euklid,
daß dreiseitige Pyramiden von gleicher Höhe sich wie ihre Grundflächen ver-
halten. Nun ist die analoge Aufgabe in der Ebene volkommen erledigt wor-
den; auch ist es Gerling gelungen, die Volumengleichheit symmetrischer
Polyeder durch Zerlegung in kongruente Teile zu beweisen. Dennoch erscheint
mir der Beweis des eben genannten Satzes von Euklid auf diese Weise im
allgemeinen wohl nicht als möglich, und es würde sich also um den strengen
Unmöglichkeitsbeweis handeln. Ein solcher wäre erbracht, sobald es gelingt,
*zwei Tetraeder mit gleicher Grundfläche und von gleicher Höhe anzugeben,
die sich auf keine Weise in kongruente Tetraeder zerlegen lassen, und die
sich auch durch Hinzufügung kongruenter Tetreader nicht zu solchen Poly-
edern ergänzen lassen, für die ihrerseits eine Zerlegung in kongruente
Tetraeder möglich ist.*"

Already in 1900 Dehn [33] gave a solution of this problem. He gives necessary conditions for polyhedrons for being zerlegungsgleich and then shows that there exist polyhedrons with equal volume that don't satisfy these conditions[8]. Only in recent times it
was shown by Sydler [105], Jessen [61] that these conditions are also sufficient.

For more literature on this subject which is more in the domain of the foundations

---

[8] This subject is related to the so called paradoxical decompositions of sets and spaces and to fundamental problems in measure theory. The surface of a sphere S^2 in $\mathbf{R}^3$ admits a decomposition into
four disjoint sets A, B, C, D, where D is denumerable, such that A, B, C and $B \cup C$ are mutually
congruent. See Hausdorff [56]. There is a vast literature on this subject.

of geometry than in that of the history of integral calculus, see Enriques [41], Hadwiger [55][9]).

1.2 *Seventeenth century*

I will pass over several centuries, to begin again in the seventeenth century. This omission is justified by the fact that in the mediaeval period mathematics had a strong algebraic character under influence of the Arabs. There were for instance the problems of the solution of algebraic equations and Diophantine equations with which the mathematicians were occupied. The theory of ratio and proportion, was, from the Greek mathematicians onward, an important method and a subject of study. But the concept of functional relation had not yet arisen. It developed very gradually in this period.

In a survey like this, it is necessary to say something about the history of the concept of function because of its close connection with our subject. On several occasions this will be evident.

The mathematicians of Antiquity did not possess our concept of function[10]). We mentioned before that Archimedes had found the quadrature of a segment of a parabola. It must be kept in mind that the parabola was defined in terms of its geometrical properties as a conic section and not, as we do nowadays, by means of an equation. This lasted for centuries. In the mediaeval period there appeared gradually an algebra of relations, for instance of the type $y = kx^n$. But it would be a misunderstanding to think that such a relation was formulated in this way: it was expressed verbally, for instance by a phrase: "one quantity is in triple ratio or proportion to another". These were *curves*, defined by geometrical properties, not functions as we say now. And for centuries, what we now call a definite integral, was an area, determined by curves or straight lines. Verbal formulations in terms of proportionalities, have also lasted for centuries. Even Leibniz used them.

The concept of relationship between variables developed gradually. It was DESCARTES (1596-1650) who, with his application of algebraical methods to geometry, opened the way for the introduction of the concept of function. Relations could now be formulated in terms of equations, gradually leading to our modern concepts. Later on there will again be occasion for some remarks on this development.

The seventeenth century is the century in which differential and integral calculus were introduced. I have not the intention to give a survey of the development of the calculus;

---

[9]) Among the mathematicians who worked on this problem I mention Wolfgang and Johann Bolyai (see [32]) and Henri Lebesgue (see [73]).

[10]) I can only give some short indications. A detailed exposition must be reserved to the historian (see for instance Boyer [22]).

this must be a subject for the historian and, moreover, there is a vast literature on these questions. Several mathematicians anticipated the calculus. PASCAL, FERMAT, WALLIS, HUYGENS have done important research on calculus in general, but their work was not in the first place pioneering as regards the notion of the integral. As to the integral KEPLER and CAVALIERI must be mentioned. I am not going to give a survey of their results. I make an exception for CAVALIERI (1598-1647), a pupil of GALILEI, who had much influence on the further development. He founded his method for calculating areas on the method of the indivisibles. In his book "Geometria indivisibilibus continuorum nova quadam ratione promota" (1635) Cavalieri develops a beginning of an integral calculus, that is to say a method for computing areas. His idea is that an area is composed of parallel line segments, a solid is to be conceived as a summation of parallel planes. In order to calculate an area, the sum of these line segments had to be determined. This concept led him to the so called *Principle of Cavalieri*, which is the following:

"*If two solids have equal altitudes, and if sections made by planes parallel to the bases and at equal distances from them are always in a given ratio, then the volumes of the solids are also in this ratio*".

With the aid of this principle Cavalieri obtained results which are equivalent to

$$_0\!\int^a x^n \mathrm{d}x = \frac{a^{n+1}}{n+1}$$

for the integers $1 \leqslant n \leqslant 9$.

He makes use of formulas comparable to

$$_0\!\int^{2a} x^n \mathrm{d}x = _0\!\int^a \left[(a+x)^n + (a-x)^n \right] \mathrm{d}x,$$

and then makes his calculations with algebraic methods and recurrence relations.

What is meant here by the words equivalent and comparable requires some explanation. Cavalieri did not possess our notation of the integral nor our concept of integral. Neither had he the concept of x^n as a function of x. In calculating what is now written as $\int x^2 \mathrm{d}x$, Cavalieri intended to compute the sum of the squares of all line segments of a certain kind in a parallelogram. And, in the same way, he had to calculate the sum of the third powers for computing $\int x^3 \mathrm{d}x$. As has already been said, he performed this in an algebraic way, in agreement with his concept of the indivisibles[11]).

For further information about Cavalieri I refer to Boyer [21].

---

[11]) A statement like "Cavalieri possessed the equivalent of the formula

$$_0\!\int^a x^2 \mathrm{d}x = \tfrac{1}{3}a^3"$$

may be dangerous when no explanation about its meaning is given. An addition "transcribed in our notation" (as is often done) is an insufficient explanation. Statements like this can give the

1.3 *Leibniz and Newton*

Based on the work of their predecessors, mentioned before, LEIBNIZ and NEWTON arrived at the introduction of differential and integral calculus. Newton (1642-1727) developed his calculus of fluxions; his notation $\dot{x}$ for $\dfrac{\mathrm{d}x}{\mathrm{d}t}$ is well known. Leibniz (1646-1716) introduced the differentials (tangents of a curve) and furthers the differential and integral calculus by introducing the algorithmic treatment and the symbols which, though slightly modified, are still in use in modern mathematics. There is no need to write about these questions, nor about the question of priority between Leibniz and Newton[12]). I only remark that the development of the integral calculus began about two thousand years before the differential calculus for the first theories which can be considered forerunners of the differential calculus were the methods developed in the seventeenth century to determine tangents on curves.

As concerns the infinitesimals of Leibniz, it is interesting to note that for some years ROBINSON and LUXEMBURG succeeded in developing a theory of infinitesimals and infinitely large numbers; they apply them to analysis[13]). This is the so called Non-standard Analysis. I give a short review.

Non-standard Analysis. Robinson [97] incorporates the theory of infinitesimals in model theory, using extensively mathematical logic; this method is too complicated to review in a few lines. There is another approach by Luxemburg.

Let $\mathbf{R}$ be the field of real numbers and $\mathbf{N}$ the set of the integers $\geqslant 0$. Let $\mathbf{R}^{\mathbf{N}}$ be the set of all mappings of $\mathbf{N}$ into $\mathbf{R}$. Defining addition and multiplication as usual $\mathbf{R}^{\mathbf{N}}$ is a ring. Let $\mathfrak{U}$ be a free ultrafilter on $\mathbf{N}$[13a]). For $A, B \in \mathbf{R}^{\mathbf{N}}$ we define $A =_{\mathfrak{U}} B$ if $\{n \mid A(n) = B(n)\} \in \mathfrak{U}$. It is shown that the relation $=_{\mathfrak{U}}$ is an equivalence relation. Denote

---

impression that Cavalieri was with his concepts much closer to us than really is the case. In interpreting historical mathematical facts one must be cautious with placing them in the framework of our present-day knowledge and with translating the facts in our language.

[12]) For this question see Fleckenstein [42].

[13]) It is interesting to read that Weyl [116] had the opinion that an analysis in fields which don't satisfy the axiom of Archimedes-Eudoxos is not useful ("Aber zugleich sieht man, dasz eine solche für die Analysis ganz unbrauchbar ist"). We must say now that this opinion was erroneous.

[13a]) A family $\mathscr{F}$ of subsets of a set X is called a filter when the following conditions are satisfied:
  (i) If $U \in \mathscr{F}$, $V \subset X$ and $V \supset U$ then $V \in \mathscr{F}$.
  (ii) The intersection of any finite number of sets of $\mathscr{F}$ belongs to $\mathscr{F}$.
  (iii) The empty set of X is not an element of $\mathscr{F}$.
  A filter $\mathscr{F}$ is called free if
  $\cap \{U \mid U \in \mathscr{F}\} = \varnothing$.
  A filter is called an ultrafilter if it is not properly contained in any other filter.
  One shows the existence of free ultrafilters on an infinite set with the aid of the axiom of choice.

by **R*** the set of the equivalence classes with respect to $=_\mathfrak{A}$ There is an injection of **R** into **R***; thus, one can suppose **R** to be embedded in **R***. An element of **R*** which belongs to **R** is called a standard element; the other elements are non-standard. **R*** is a totally ordered field, not isomorphic to **R**; the order is non-archimedean. Now, in **R*** one can define infinitesimals and infinitely large numbers. The infinitesimals have the properties that Leibniz attributed to them. For an exposition of Luxemburg's method see [79].

For a description of the history of infinitesimals see Robinson [97].

I now come to the theory of the integration. In the seventeenth century there occurs a transformation of the problem; one is no longer interested so much in the determination of areas, rather one studies relations between curves (later: between functions) to which the study of areas gave rise. In this context integration appears as the inverse of differentiation. In modern formulation:

Let a function f be given; one asks to find all functions F such that F′=f.

This is the problem of the existence of primitive functions.

For certain classes of functions this problem was solved in relation with the concept of area in a well known way.

Let C be the curve $y=f(x)$. Denote by $S(x)$ the area of the figure, limited by the straight lines $x=a$ and $x=x_0$. Then

$$S'(x)=f(x),$$

and

$$S(x)={_a}\!\int^x f(t)\,\mathrm{d}t.$$

This was proved by considering the area, included by the lines $x=x_0$ and $x=x_0+\varDelta x_0$. Some explanation is necessary.

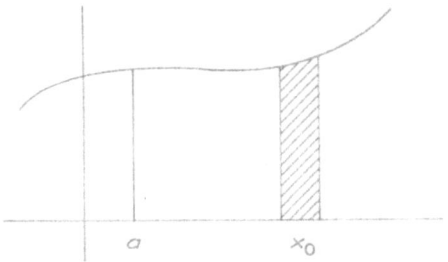

First, Newton formulated the problem as finding the "fluent" if the "fluxion" is given. He did not use the notation $\int f(t)\,\mathrm{d}t$, which was introduced by Leibniz. The integral sign $\int$ developed from Leibniz' use of the long letter s (abbreviation of "summa"), in normal type $\int$, in italics $\int$. Leibniz used the word "summa"; the word "integral" appears for the first time in the work of the Bernoulli's. I remark that integration was not considered as an operator on the function f.

In the second place I make again some remarks on the notion of function. Discussions about this notion can be found up to the beginning of our century. I have the intention to return to this subject in the paragraph on Borel and Lebesgue. Here I make some general remarks on the development of the notion of function in the years after Leibniz and Newton.

First about the introduction of the *word* "function". This word appears for the first time in a paper of Leibniz in the Acta Eruditorum of 1692, but it is used there in a very special way, namely to design geometrical quantities (subtangents, subnormals and so on) depending on a point of a curve. In 1698 Johann Bernoulli I uses the word, writing "Funktionen der Ordinaten". There is a definition of Johan Bernoulli I from 1718: "On appelle fonction d'une grandeur variable une quantité composée quelque manière que ce soit de cette grandeur variable et de constantes". See Tropfke [107] Bd II, p. 35, 36. I mention the the notation $f(x)$, $\varphi(x)$ (nowadays by well known arguments f, φ) is due to EULER (1707-1783).

The idea of what a function is, differed in these years considerably from ours.

A function was a relation between two variables which could be described by one law, which was defined by the usual operations of addition, multiplication, trigonometrical and logarithmic operations and so on. A function S of x as before (the area), was considered to be of a different kind, namely determined by a geometrical relation. One distinguished "genuine" functions and those which were not "genuine". Continuous functions were functions of the first kind. It was believed that any continuous function, whether or not defined in a geometrical way, could be represented by an analytical expression and that this was not the case for non-continuous functions.

In the course of the development of analysis it became clear that this concept was not the right one. The theory of Fourier series furnished examples of discontinuous functions, which can be represented by a convergent trigonometrical series, for example the function, defined on $[0,2\pi]$, being 0 on $[0,\pi]$ and 1 on $[\pi,2\pi]$[14]).

DIRICHLET (1805-1859) gave the following example of an extremely discontinuous function which nevertheless has an analytic representation: it is the function

$$f : f(x) = \lim_{m \to \infty} \left[\lim_{n \to \infty} (\cos(m!\pi x))^{2n} \right]$$

which equals 0 for all irrational values of x and equals 1 for all rational values of x.

---

[14]) There has always been a close relation between the theory of trigonometrical series and the theory of real functions and integration theory. For an account of these relations see a paper by Paplauskas [89].

Even Dirichlet, who lived in the same period as Cauchy – the first mathematician who gave an exact definition of continuity – gave a verbal description of continuity. In an important paper he wrote as follows[15]):

" ÜBER DIE DARSTELLUNG GANZ WILLKÜRLICHER FUNCTIONEN DURCH SINUS- UND COSINUSREIHEN.

Die merkwürdigen Reihen, welche in einem bestimmten Intervalle Functionen darstellen, welche ganz gesetzlos sind oder in verschiedenen Theilen dieses Intervalls ganz verschiedenen Gesetzen folgen, haben seit der Begründung der mathematischen Wärmelehre durch Fourier so zahlreiche Anwendungen in der analytischen Behandlung physikalischer Probleme gefunden, dass es zweckmässig erscheint, die für die folgenden Bände dieses Werkes bestimmten Auszüge aus den neuesten Arbeiten über einige Theile der mathematischen Physik durch die Entwickelung einiger der wichtigsten dieser Reihen einzuleiten.

§. 1.

Man denke sich unter a und b zwei feste Werthe und unter x eine veränderliche Grösse, welche nach und nach alle zwischen a und b liegenden Werthe annehmen soll. Entspricht nun jedem x ein einziges, endliches y, und zwar so, dass, während x das Intervall von a bis b stetig durchläuft, $y = f(x)$ sich ebenfalls allmählich verändert, so heisst y eine stetige oder continuirliche Funktion von x für dieses Intervall. Es ist dabei gar nicht nöthig, dass y in diesem ganzen Intervalle nach demselben Gesetze von x abhängig sei, ja man braucht nicht einmal an eine durch mathematische Operationen ausdrückbare Abhängigkeit zu denken. Geometrisch dargestellt, d.h. x und y als Abscisse und Ordinate gedacht, erscheint eine stetige Function als eine zusammenhängende Curve, von der jeder zwischen a und b enthaltenen Abscisse nur ein Punkt entspricht. Diese Definition schreibt den einzelnen Theilen der Curve kein gemeinsames Gesetz vor; man kann sich dieselbe aus den verschiedenartigsten Theilen zusammengesetzt oder ganz gesetzlos gezeichnet denken. Es geht hieraus hervor, dass eine solche Function für ein Intervall als vollständig bestimmt nur dann anzusehen ist, wenn sie entweder für den ganzen Umfang desselben graphisch gegeben ist, oder mathematischen, für die einzelnen Theile desselben geltenden Gesetzen unterworfen wird. So lange man über eine Function nur für einen Theil des Intervalls bestimmt hat, bleibt die Art ihrer Fortsetzung für das übrige ganz der Willkür überlassen."

---

[15]) Repertorium der Physik, Bd. I (1837), 152-174. See [66].

A. F. Monna

In a note "Mémoire sur les fonctions continues" CAUCHY (1789-1857) criticizes the situation in the following way ([28])[16]).

> "Dans les ouvrages d'Euler et de Lagrange, une fonction est appellée continue ou discontinue, suivant que les diverses valeurs de cette fonction, correspondantes à diverses valeurs de la variable, sont ou ne sont pas assujetties à une même loi, sont ou ne sont pas fournie par une seule et même équation."

Cauchy observes that this definition lacks mathematical exactness. He gives in the same note the following example.

> "Ainsi, par exemple, x désignant une variable réelle, une fonction qui se réduirait, tantôt à $+x$, tantôt à $-x$, suivant que la variable x serait positive ou négative, devra, pour ce motif, être rangée dans la classe des fonctions discontinues, et cependant la même fonction pourra être regardée comme continue, quand on la représente par l'intégrale définie
>
> $$\frac{2}{\pi} \int_0^\infty \frac{x^2 \mathrm{d}t}{t^2 + x^2}$$
>
> ou même par le radical
>
> $$\sqrt{x^2},$$
>
> qui est la valeur particulière de la fonction continue $\sqrt{x^2 + t^2}$, correspondante à une valeur nulle de t. Ainsi, le caractère de continuité dans les fonctions, envisagé sous le point de vue auquel se sont d'abord arrêtés les geomètres, est un caractère vague et indéterminé. Mais l'indétermination cessera si à la définition d'Euler on substitue celle que j'ai donné dans le Chapitre II de l'Analyse algébrique."[17])

Cauchy has given a definition of continuity in the sense as we do now[18]):
"... la valeur numérique de la différence $f(x+\alpha) - f(x)$ décroit indéfiniment avec celle de α. En d'autres termes, la fonction $f(x)$ restera continue par

---

[16]) See Cauchy, Oeuvres complètes 1e série, t. VIII, p. 145.
[17]) Mr. van Dormolen informs me that he has the experience that pupils of secondary schools would even today criticize the function of the example of Cauchy by saying: it is not fair to define a function in that way.
[18]) Cours d'Analyse de l'école royale polytechnique (1821). Oeuvres complètes 2e série t. III, p. 43.

90

rapport à x entre les limites données si, entre ces limites, un accroissement infiniment petit de la variable produit toujours un accroissement infiniment petit de la fonction elle même."

He still uses infinitesimals which he defines as follows[19]):

"Lorsque les valeurs numériques successives d'une même variable décroissent indéfiniment de manière à s'abaisser au dessous de tout nombre donné, cette variable devient ce qu'on nomme un infiniment petit ou une quantité infiniment petite. Une variable de cette espèce a zéro pour limite."

However, Cauchy did not use explicitly his definition of continuity; there was no need to consider functions which are continuous only because all continuous functions were as a matter of course considered differentiable. This point of view changed after Weierstrass had given in his courses an example of a nowhere differentiable continuous function (1861)[20]).

Cauchy was the first mathematician who defined an integral for continuous functions in an analytical way by means of finite partitions of the given interval, thus anticipating the Riemann-sums. He shows in this way the existence of primitive functions (the area-function S mentioned before). He is vague in his conclusions; for instance: "Donc, lorsque les éléments de la différence $X - x_0$ deviennent infiniment petits, le mode de division n'a plus sur la valeur de S qu'une influence insensible ..." (l.c. 19) p. 125).
Cauchy gives the notations

$$\int_{x_0}^{x} f(x)\,\mathrm{d}x, \quad \int f(x)\,\mathrm{d}x \begin{bmatrix} x_0 \\ x \end{bmatrix}, \quad \int f(x)\,\mathrm{d}x \begin{bmatrix} x = x_0 \\ x = X \end{bmatrix}$$

and he observes that the first, which was proposed by Fournier, is the most simple.

2 Riemann, Lebesgue, real functions

With Riemann a new period in the development of the integral begins. He is the first

---

[19]) Résumé des leçons données à l'école royale polytechnique sur le calcul infinitésimal (1823), Oeuvres complètes, 2e série t. IV.

[20]) In the course of the years many examples were given. Lebesgue [74], trying to find an example fit for elementary courses, gave also a simple example. See Enzyklopädie der Math. Wissenschaften II A2, no. 4 (A. Voss).

BERNHARD RIEMANN

of a series of famous mathematicians who have gradually built our modern theory of the integral: Jordan, Borel, Lebesgue, Denjoy.

A line can be drawn from the Greek mathematicians to their theories, this means that there is a strong relation to area and volume: now in its modern form to measure theory. It is only in recent times that the functional analytic character of the integral dominates: there the integral is considered as a mapping.

There is another domain in mathematics which I shall have to consider, it is the analysis of real functions. In the years round 1900 the theory of the integral and the theory of real functions were connected with each other in such a way, that they can scarcely be separated from each other in a review.

2.1 *Riemann*

RIEMANN (1826-1866) began his studies in 1846 in Göttingen, first in theology, but he turned soon to mathematics; at that time Gauss was in Göttingen.

1847-1849   Mathematical studies in Berlin (Jacobi, Lejeune Dirichlet, Steiner).

1849   Back to Göttingen.

1851   Dissertation in Göttingen "Grundlagen für eine allgemeine Theorie der Functionen einer veränderlichen complexen Grösse."

1853 Habilitationsschrift in Göttingen "Ueber die Darstellbarkeit einer Function durch eine trigonometrische Reihe".
To accomplish his "Habilitation" Riemann had to give a lecture. From the three subjects, proposed by Riemann, Gauss chose "Ueber die Hypothesen welche der Geometrie zu Grunde liegen"; it became a famous paper.
1855 After the death of Gauss, Dirichlet succeeded Gauss in Göttingen.
1857 Riemann professor extraordinary in Göttingen.
1859 After the death of Dirichlet, Riemann succeeded Dirichlet in Göttingen.

The important influence of Riemann on the development of mathematics is too large to be reviewed in a few lines, and after all this is well known by every mathematician. For a biography of Riemann see [95].

Cauchy defined the integral for continuous functions (there are some generalizations when the function f is not bounded; improper integrals). Riemann proceeds in a more general way; he puts the question: what should be the meaning of $_a\int^b f(x)dx$. For this he considers finite partitions of $[a,b]$ and forms the well known sum (in Riemann's notation)

$$S = \sum_{i=1}^{n} \delta_i f(x_{i-1} + \varepsilon_i \delta_i).$$

When S converges for all $0 \leqslant \varepsilon_i \leqslant 1$ if all δ_i tend to 0 ("sobald sämtliche δ unendlich klein werden"), then the limit is called by Riemann the integral $_a\int^b f(x)dx$. I observe that the limit of the sums must be defined in an appropriate way, namely by means of the filter of the finite partitions of $[a, b]$. Not restricting himself to continuous functions, Riemann analyzes the question under what conditions this limit exists. He gives necessary and sufficient conditions:

> "Damit die Summe S, wenn sämtliche δ unendlich klein werden, convergirt, ist ausser der Endlichkeit der Function $f(x)$, noch erforderlich, dass die Gesammtgrösse der Intervalle, in welchen die Schwankungen $> \sigma$ sind, was auch σ sei, durch geeignete Wahl von δ beliebig klein gemacht werden kann."

He shows that this condition is also sufficient[21].
Anticipating measure theory, this criterion is in modern terms[22]:

---

[21] Riemann gave his definition in a paper "Über die Darstellbarkeit einer Function durch eine trigonometrische Reihe", composed in 1854, but published after his death in 1867; see the note of Dedekind in the collected works of Riemann [95]. The paper contains a historical introduction concerning the representation of a function by means of trigonometrical series.

[22] The criterion, using the notion of a set of measure zero, could be formulated in this way more than 40 years later when Borel had given his theory.

In order that the bounded function f is integrable (in the sense of Riemann) it is necessary and sufficient that the set of the points where f is not continuous has measure 0.

Riemann did not define a lower and an upper integral. They were introduced by DARBOUX in 1875 in a paper "Mémoire sur les fonctions discontinues", in which he uses Riemann's integral for constructing continuous functions without derivative. Because of the importance of this question for the development of analysis, I quote here the introduction of this paper (see [31a]).

"Jusqu'à l'apparition du Mémoire de Riemann sur les séries trigonométriques aucun doute ne s'était élevé sur l'existence de la dérivée des fonctions continues. D'excellents, d'illustres géomètres, au nombre desquels il faut compter Ampère, avaient essayé des démonstrations rigoureuses de l'existence de la dérivée. Ces tentatives étaient loin sans doute d'être satisfaisante; mais, je le répète, aucun doute n'avait été formulé sur l'existence même d'une dérivée pour les fonctions continues.

La publication de Riemann a décidé la question en sens contraire."[23])

2.2 *Measure theory*

At the base of Riemann's definition of the integral is the ordinary length of an interval in **R**. This fact, and Riemann's criterion formulated in the form as Riemann did, make it evident that, in considering generalizations and improvements of Riemann's integral, we shall have to speak first about the development of *measure theory*.

For the moment I indicate that measure theory is concerned with finding generalizations of the notions of length, area and volume to a larger class of sets, preserving the known properties of additivity and continuity. This will be clear from the examples, given in the course of the book. Furtheron (page 100) I shall give a modern definition of the notion of a measure, defined on a set X.

In our formulation of Riemann's criterion sets of measure 0 are introduced. Here, a set $A \subset \mathbf{R}$ is said to have measure 0 if, for any $\varepsilon > 0$, A can be included in the union of at most enumerable intervals with total length $\leqslant \varepsilon$. In the course of the development of measure theory, the enumerability of the set of intervals is of essential importance.

CANTOR was one of the first[24]) who gave a general definition of the measure of a set.

---

[23]) The main work of the French mathematician Darboux is not in this domain. He is known by his famous book on differential geometry: Leçons sur la théorie générale des surfaces et les applications géométriques du calcul infinitésimal (I, II, III, IV) (Paris 1887).

[24]) About the same time are definitions of Stolz and Harnack; see Enzyklopädie der Math. Wissenschaft, II C9a (Zoretti-Rosenthal).

He introduced it in 1884 in the course of his development of set theory (see [25]) by the following method.

Let A be a bounded set in $\mathbf{R}^n$; for any $\rho > 0$ put

$$B_{x,\rho} = \{\xi \in \mathbf{R}^n \mid \|\xi - x\| \leqslant \rho\},$$

$$W_\rho = \bigcup_{x \in A} B_{x,\rho}.$$

Cantor remarks that the volume $m(W_\rho)$ of W_ρ can be calculated by means of a computation of an integral (but he is not clear in showing how). The measure $m(A)$ of A is then defined to be

$$m(A) = \lim_{\rho \to 0} m(W_\rho).$$

Thus every bounded set has a measure. Cantor observes that a set and its closure have the same measure. It is a consequence of this fact that this measure is not additive. If, for instance, A is the set of rational numbers in $I = [0,1]$, B is the set of irrational numbers in $[0,1]$, then $m(A) = m(B) = 1$ and $m(A \cup B) = m(I) = 1$. Cantor did not mention this consequence; perhaps the conclusion is allowed that Cantor did not have the idea that additivity is a condition to be put on a measure. It is just this condition which is important in the development of measure and integral. Cantor remarks with emphasis that the value of the measure, defined in this way, depends on the dimension of the space in which the set is embedded[25]).

The next step in the development of measure theory was done by the French mathematician C. Jordan[26]).

CAMILLE JORDAN (1838-1922) was born in Lyon. He began his career as an engineer but he turned to mathematics and in 1873 he accepted a position at the Ecole polytechnique in Paris, where he became a professor in 1876, succeeding Hermite. From 1883 on he was also professor at the Collège de France, succeeding there Liouville. In 1912 he retired. Hadamard succeeded him at the Ecole polytechnique and Humbert at the Collège de France. Jordan is especially famous for his work on the theory of groups; we mention his "Traité des substitutions et des équations algébriques". But he published also on the theory of measure and on real functions His work on Analysis situs (topology) must be mentioned.

For a biography of Jordan, see Lebesgue [75].

---

[25]) Cantor states that this general notion of volume of an arbitrary set is indispensible for him in his research on the dimension of sets. There is indeed in modern dimension theory a connection between the concepts of measure and dimension; see Nagata [87], Bouligand [16].

[26]) For the theory of measure in the initial period see also Peano [90].

A. F. Monna

In the theory of measure Jordan proceeds as follows. Let V be a bounded set in **R**. Let (I_n) be any finite system of closed intervals covering V. The measure $m(V)$ of V is defined the lower bound of the numbers $\sum_n |I_n|$, where $|I_n|$ is the length of the interval I_n. If $\overline{V}$ denotes the closure of V, one has $m(V) = m(\overline{V})$. Later the number $m(V)$, defined in this way, is called the *outer measure* of V and it is then denoted by $\overline{m}(V)$. An *inner measure* is defined by

$$\underline{m}(V) = |I| - \overline{m}(I - V),$$

where I is a closed interval containing V. Now V is called *measurable (Jordan-measurable)* if

$$\underline{m}(V) = \overline{m}(V).$$

There are several objections to this concept of measure:
(a)   There are open sets which are not measurable;
(b)   The set of rational numbers in an interval I is not measurable;
(c)   The measure is additive only in a restricted sense, that is to say if (V_i) is a system of measurable sets, $V_i \cap V_j = \varnothing$ $(i \neq j)$, $V = \cup V_i$ then

$$m(V) = \sum_i m(V_i)$$

holds only for finite systems, and not for enumerable systems.

96

For this theory of Jordan see his *Cours d'analyse* de l'école polytechnique I, second edition (1893), where it is incorpcrated in a paragraph on set theory. Note that in the first edition of this Cours d'analyse from 1882 set theory is not yet treated. In his work on analysis Jordan still uses the terminology of infinitesimals, calling "infin:men: petit" any variable which tends to 0[27]).

A theory of measure in $\mathbf{R}^n$ can be developed in a similar way, in particular in $\mathbf{R}^2$, and from this there is a way to the Riemann integral of a function f in the sense that $_a\int^b f(x)\mathrm{d}x$ is equal to the Jordan measure of the set of all ordinates corresponding with f, when this set is measurable. If $A \subset \mathbf{R}$ is Jordan measurable and if φ denotes the characteristic function of A (that is $\varphi(x)=1$ for $x \in A$, $\varphi(x)=0$ for $x \notin A$) then

$$m(A)=\int \varphi \, \mathrm{d}x.$$

An important contribution to measure theory was given by Emile Borel in 1398 ([9]).

EMILE BOREL (1871-1956) was born in Saint-Affrique (Provence). His dissertation (Sur quelques points de la théorie des fonctions) dates from 1892.

1893 Maître de conférences à la Faculté des Sciences de Lille.

1897 Maître de conférences at the Ecole normale supérieure in Paris.

1909 The Faculté des Sciences in Paris creates a new chair on "Theorie des fonctions" and Borel is the first professor in this chair.

1920 Borel changes his chair for the one on "Théorie des probabilités et Physique mathématique", which was formerly the chair of Poincaré.

Borel started his mathematical work on analysis, but gradually his interest turned to probability and mathematical physics, especially after the war 1914-1918. He wrote more than 300 books and articles in several domains: pure mathematics, popular books ("La nouvelle collection scientifique"), physics and the application of mathematics. He published some articles on the theory of games and econometrics. As to the last subjects there was a question of priority between him and J. von Neumann.

Borel had an active part in the creation of the "Institut Henri Poincaré" (1928). He was interested in teaching and wrote some books for secondary schools.

Borel became interested in politics under the influence of the mathematician Paul Painlevé, who was Prime Minister about 1915; Borel was for some time Minister under Painlevé. He was a member of Parliament for years.

For a biography of Borel see [47], [30].

---

[27] As an illustration of the importance of Jordan the following fact is noteworthy. Since 1885, Jordan was charged with the direction of the "Journal de Mathématiques Pures et Applicuées", which was founded in 1836 by Liouville. Since, this journal was often referred to by f-ench mathematicians as "Le journal de M. Jordan" (see for instance Borel [13]).

EMILE BOREL

Borel opened the way to Lebesgue's definition of an integral.

The concept of Borel-measure is based on the property that any open set $V \subset \mathbf{R}$ is the union of countably many disjoint open intervals:

$$V = \bigcup_{i=1}^{\infty} I_i.$$

For a bounded open set V, the measure is defined by

$$m(V) = \sum_{i=1}^{\infty} |I_i|,$$

which is finite, V being bounded[28]).

One proves that for any open bounded set this measure is uniquely determined, that is to say if there is more than one decomposition of V, the method gives the same value

---

[28]) The notations $A \cup B$ and $A \cap B$ respectively for the union and the intersection of the sets A and B were introduced by Peano in an introductory chapter on the operations in logic in his book "Calcolo geometrico secondo l'Ausdehnungslehre di H. Grassmann" (Turijn 1888). Sometimes he writes AB instead of $A \cap B$, calling this the product of A and B. However, the mathematicians did not follow Peano in this notation. In the older literature the union of the sets A and B is called the sum and is denoted by $A + B$. The intersection is called the product and denoted by $A.B$ or AB. Later on one finds again $A \cup B$ and $A \cap B$.

RENÉ BAIRE

in each case. This justifies saying that bounded open sets are *measurable (Borel-measurable)*.

It is proved that this measure – defined for open bounded sets – is completely additive (nowadays called σ-additive), that is:

If $V = \bigcup_{i=1}^{\infty} V_i$, V and (V_i) measurable, the sets V_i being disjoint, then

$$m(V) = \sum_{i=1}^{\infty} m(V_i).$$

Borel introduced this measure in his book from 1898 [9], however not exactly in this form because he did not yet use the terminology of open sets.

Open sets.

The concept of an open set as well as the denomination "ensemble ouvert" is presumably introduced by R. BAIRE in his thesis "Sur les fonctions de variables réelles" [2] in 1899. He defined an open sphere and an open set in exactly the same way as we do now in metric spaces ("domaine ouvert à n dimensions"). Without reference to Baire, Lebesgue writes in his thesis ([69], p. 242):

"On peut encore dire que l'étendu extérieure de E est la mesure de l'ensemble de ses points intérieurs, lequel ensemble étant ouvert (*), c'est à dire ne contenant aucun point de sa frontière, a pour complémentaire un ensemble fermé et par suite est mesurable (*B*)".

The footnote (*) reads like this:
"Tous les points d'un tel ensemble sont intérieurs a l'ensemble".

Schlesinger and Plessner [99], p. 15, gave the following definition "Eine Punktmenge, die aus lauter inneren Punkte besteht, heiszt offen", with reference to the passage in Lebesgues thesis which I just quoted.

Neither with Cantor, nor with Borel (1898) could I find the concept of open set. This is not strange, because in the first decades of the theory of sets closed sets and especially perfect sets were investigated by the mathematicians, not the sets which we now call open. This strikes one when reading books and papers written in these years. In contemporary mathematics, on the contrary, open sets seem to be more fundamental than closed sets.

The measure can be extended in a well known way to a wider class of sets. If $I = [a, b]$ and F is closed $F \subset I$, then the complement O of F relative to I is open in I and the measure $m(F)$ of F is defined by

$$m(F) = b - a - m(O).$$

Transfinitely repeated application of the set-theoretic operations of taking countable unions and complements leads then to a family of sets for which a measure is defined. They are the so called *Borel-sets* and are called *B-measurable*.

This method is important in modern measure theory, but it is introduced there in an axiomatic form. One defines then the concepts of σ-ring and σ-algebra.

Let X be a non-empty point set. A non-empty collection Γ of subsets of X is called a σ-ring if

(i) $V_n \in \Gamma \, (n = 1, 2, \ldots) \Rightarrow \bigcup_n V_n \in \Gamma$;

(ii) $V \in \Gamma, W \in \Gamma \Rightarrow V - W \in \Gamma.$

A σ-ring Γ is called a σ-algebra (or σ-field) if $X \in \Gamma$. A measure on Γ is a mapping m from Γ into the set of non-negative real numbers satisfying

$$m \left(\bigcup_{i=1}^{\infty} V_i \right) = \sum_{i=1}^{\infty} m(V_i)$$

for any family of disjoint sets $V_i \in \Gamma$.

2.3 *Lebesgue*

HENRI LEBESGUE (1875-1941) studied at the Ecole normale supérieure in Paris. For a short time he was a teacher at a lycée in Nancy. Untill 1906 he was Maître de conférences in Rennes. In that year he went to Poitiers, "chargé de cours", and later professor. In 1912 he went to Paris as Maître de conférences, where he became professor at the Collège de France.

Information on the scientific work of Lebesgue will be found in many places in this book.

Borel did not use his theory of measure for defining a new concept of integral. This was done by Lebesgue in his thesis: "Intégrale, longueur, aire" in 1902 [69]. See also Lebesgue's famous book "Leçons sur l'intégration et la recherche des fonctions primitives", first edition 1903 ([70]), second edition 1926.

Lebesgue develops the theory of measure in a more general form than Borel; he assigns an inner and an outer measure to every set. Let V be a set. Then the *outer measure* $\bar{m}(V)$ is defined by

$$\bar{m}(V) = \inf m(O),$$

the infimum taken over all open sets O containing V. The inner measure is defined to be

$$m(V) = \sup m(F)$$

over all closed sets $F \subset V$.

A set is called measurable if $\bar{m}(V) = m(V)$. This measure m is completely additive; i.e. the additivity holds for enumerable families.

Lebesgue defines a measure in $\mathbf{R}^2$ and this led him to a *geometrical definition* of the integral of a positive function f, defined for $a \leqslant x \leqslant b$. It is the two dimensional measure of the set of the ordinates $0 \leqslant y \leqslant f(x)$, $a \leqslant x \leqslant b$ when this measure exists. This is generalized for functions assuming both signs.

But Lebesgue gave also an interesting *analytical method* for defining the integral; his method differs essentially from that of Riemann.

Let f be a bounded real function, defined on $[a,b]$. Suppose $m \leqslant f(x) \leqslant M$ for $a \leqslant x \leqslant b$. For any $\xi \leqslant \eta$, define

$$V_{\xi,\eta} = \{x \in [a,b] \mid \xi \leqslant f(x) < \eta\}.$$

Suppose $V_{\xi,\eta}$ is measurable for all ξ and η; the function f is then called measurable.

The difference with Riemann's method is this, that Lebesgue does not consider partitions of the domain $[a,b]$ of f, but of the range of f.

Consider the partition

$$m = \xi_1 < \xi_2 < \ldots < \xi_n < \xi_{n+1} = M,$$

and put

$$V_i = \{x \mid \xi_i \leqslant f(x) < \xi_{i+1}\}.$$

Lebesgue considers the sums

$$\sum_{i=1}^{n} \xi_i \, m(V_i), \quad \sum_{i=1}^{n} \xi_{i+1} \, m(V_i).$$

Proceeding as for the Riemann-integral, but now with respect to the partitions of the range of f, the integral of f is defined to be the common limit (defined in an adequate way; see p. 93) of these sums which can be proved to exist.
It is then denoted by

$$_a\!\int^b f(x) \, \mathrm{d}x \quad \text{or} \quad (\mathfrak{L}) \, _a\!\int^b f(x) \, \mathrm{d}x,$$

and called the *Lebesgue integral*.

Lebesgue himself gave the following curious description of his method comparing it to the method of Riemann; I found it in a biography of Lebesgue, written by A. Denjoy, Mme. L. Félix and P. Montel [38].

"Lebesgue expliquait la nature de son intégrale par une image plaisante et accessible à tous. "Je dois payer une certaine somme, disait-il; je fouille dans mes poches et j'en sors des pièces et des billets de différentes valeurs. Je les verse à mon créancier dans l'ordre ou elles se présentent jusqu'à atteindre le total de ma dette. C'est l'intégrale de Riemann. Mais je peux opérer autrement. Ayant sorti tout mon argent, je réunis les billets de même valeur, les pièces semblables, et j'effectue le paiement en donnant ensemble les signes monétaires de même valeur. C'est mon intégrale." "[29])

I now have to make several remarks about the properties of the integral of Lebesgue.
First some observations on the properties of the $\mathfrak{L}$-integral.

---

[29]) Defining measurability of functions with the method of Lebesgue, but then starting from Jordan-measure, does not lead to a good theory, because there are examples of continuous functions which would not be measurable.

Example. Let V be a non-dense perfect set in $I = [0,1]$, obtained in a well known way as the complement of a family of disjoint open intervals in I; the $\mathfrak{L}$-measure of V be > 0. Define a function f on I as follows. Put $f(x) = 0$ for $x \in V$. On any of the open intervals (a, β) of the family, define f

(i)   Lebesgue generalized his theory for functions of more variables and to certain classes of functions which are not bounded.

(ii)   Theorems concerning the convergence of the integral of a sequence of functions (f_n), culminating in the famous theorem on dominated convergence: if f_n is $\mathfrak{L}$-integrable and if $\lim_{n \to x} f_n(x) = f(x)$ exists almost everywhere (see p. 109) and there is a $\mathfrak{L}$-integrable function g such that $|f_n(x)| \leqslant g(x)$ almost everywhere, then f is $\mathfrak{L}$-integrable and

$$\lim_{n \to \infty} \smallint\smallint f_n \, dx = \smallint f \, dx.$$

This is one of the properties in which the Lebesgue-integral is superior to the Riemann-integral where usually this theorem is proved under the condition of uniform convergence.

(iii)   The relation between the Riemann- and the Lebesgue-integral.

If the bounded function f on $[a,b]$ is Riemann-integrable, then it is Lebesgue-integrable (summable) and the integrals are equal. There are functions which are Lebesgue-integrable but not Riemann-integrable, for instance the characteristic function of the set of rational numbers in $[0,1]$.

However, for functions which are not bounded the situation is more complicated as is shown by the following example which is to be found in the thesis of Lebesgue.

The function f defined on $[0,1]$ by

$$f(x) = \frac{d}{dx}\left(x^2 \sin \frac{1}{x^2}\right) = 2x \sin \frac{1}{x^2} - \frac{2}{x} \cos \frac{1}{x^2} \quad (x \neq 0)$$

$$f(0) = 0$$

is not Lebesgue-integrable over $[0,1]$, although it is continuous on $[\varepsilon,1]$ for any $0 < \varepsilon < 1$. Nevertheless f is integrable in the sense of Cauchy-Riemann because

$$\lim_{\varepsilon \to 0} {}_\varepsilon\!\smallint^1 f(x) \, dx$$

exists. Note that f is not bounded.

---

as in the figure, the triangles being equilateral. Thus f is continuous on I, but is not Jordan measurable.

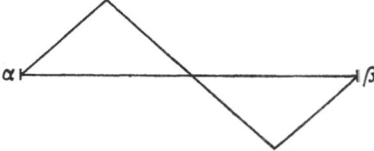

It is not possible to define the Riemann-integral by means of dividing the range of the function and with the use of Jordan-measure, because the subsets are not always measurable. See an article of Montel and Rosenthal in the Enzyklopädie der Math. Wiss. [84].

We need not be too astonished at this fact because it is a property of the $\mathfrak{L}$-integral that if a function f is $\mathfrak{L}$-integrable, then $|f|$ is $\mathfrak{L}$-integrable and in this example

$$_\varepsilon\!\int^1 |f(x)|\,dx \to \infty \text{ if } \varepsilon \downarrow 0.$$

As to the question of the existence of bounded functions which are not $\mathfrak{L}$-integrable, Lebesgue remarks that he does not know whether they exist:

> "Je ne connais aucune fonction qui ne soit sommable, je ne sais s'il en existe. Toutes les fonctions qu'on peut définir à l'aide des opérations arithmétiques et du passage à la limite sont sommables. Toutes les fonctions intégrables au sens de Riemann sont sommables et les deux définitions de l'intégrale conduisent au même nombre. Toute fonction dérivée bornée est sommable"
> (see [69]).

This problem was solved by VITALI in 1905 ([109]). He gave an example of a set which is not measurable in the sense of Lebesgue, thus answering the question in positive sense because the characteristic function of such a set is not $\mathfrak{L}$-integrable. I reproduce this example here (see Kamke [63], Zaanen [118]).

Let ξ be an irrational number or 0. Denote by $A(\xi)$ the set of all real numbers $\xi + r$, where r is any rational number. If ξ_1 and ξ_2 are irrational numbers, clearly $A(\xi_1) = A(\xi_2)$ if and only if $\xi_1 - \xi_2$ is a rational number. Let V be the set of all different sets $A(\xi)$. Choose from each of the elements $A(\xi)$ of V a number $\eta(\xi)$ such that $0 < \eta(\xi) < 1$. The set $W(\eta)$ of all these numbers η is a set which is not $\mathfrak{L}$-measurable.

Suppose $W(\eta)$ was $\mathfrak{L}$-measurable. Then all the sets $W\left(\eta + \dfrac{1}{n}\right), n = 1,2,\ldots$, are $\mathfrak{L}$-measurable. It is easily shown that these sets are disjoint, and so the union of $\overset{\infty}{\underset{n=1}{\cup}} W\left(\eta + \dfrac{1}{n}\right)$ and $W(\eta)$ is $\mathfrak{L}$-measurable and the measure is $\leqslant 1$. Because all these sets have equal measure it follows $m(W(\eta)) = 0$ and $m(W(\eta + s)) = 0$ for any rational number. Denoting by $D(\eta + s)$ the intersection of the interval $[0,1]$ with $W(\eta + s)$, it follows $m(D(\eta + s)) = 0$. The union of the sets $D(\eta + s)$, where s is any rational number, has then measure 0 and this is a contradiction because this union is the interval $[0,1]$.

Kamke (1925) remarks that this example is based on the Axiom of Choice (Auswahlprinzip) and that it is not known whether an example of a non-measurable set can be constructed without this axiom[30]).

---

[30]) To judge the value of this remark nowadays, one must take into account the modern results on the Axiom of Choice (see part 1). Assuming the existence of certain large cardinals, one can construct a model of set theory in which the Axiom of Dependent Choices (a weak form of the Axiom of

Not all mathematicians were convinced that non-measurable sets exist. The following passage, which I take from a letter from Nicolas Lusin to W. Sierpinski (18 july 1926) is very interesting (Sur une question concernant la propriété de M. Baire, Fund. Math. IX, 1927) [78a]:

"Vous me demandez mon opinion sur la question suivante: *Tout ensemble de points qui possède la propriété de M. René Baire doit-il être nécessairement mesurable?*

La voici.

Oui, puisque je ne crois nullement à l'existence d'ensembles non-mesurables. Quand j'entends parler d'une loi définissant *un* ensemble "sans propriété de M. René Baire" ou bien "non measurable" *etc.* je suis très méfiant, parce que je n'ai jamais encore vu de pareilles lois. Mais ce n'est qu'une affaire de routine et, à la réflexion, je vois déjà des difficultés aussi graves, à mon avis, dans les raisonnements où interviennent des ensembles non mesurables *B*: on ne sait pas s'il est possible de nommer un ensemble non mesurable *B* défini d'une manière *positive* et sans faire intervenir, explicitement ou *implicitement*, la notion du transfini. Ainsi, je crois que *tout* ensemble de points *réellement défini* est nécessairement mesurable et, peut-être, toujours même mesurable *B*.

Non, si nous abandonnons le terrain des réalités mathématiques pour celui des déductions logiques purement verbales."

(iv) As is seen from the title of his "Leçons", Lebesgue studied the problem of the existence of primitive functions:

(a) *characterize the functions which are the derivative of an other function,* (b) *given the derivative, find the function.*

The Riemann-integral does not give good results for this problem. For continuous functions it is easy. Let f be R-integrable over $[a,b]$. Put

$$F(x) = {}_a\!\int^x f(t)\,dt.$$

Then $F'(x) = f(x)$ in every point $x \in (a,b)$ where f is continuous. A primitive function is not uniquely determined.

But the problem is much more difficult when f is not continuous. On the one hand there are trivial examples of functions f which are R-integrable but for which F is not differentiable for certain values of x. Lebesgue gave interesting examples; I quote the following part from his "Leçons" (Chapitre V, La recherche des fonctions primitives).

Choice) holds and in which every set of reals is Lebesgue measurable. Questions like this belong to model theory and to the domain of the axiomatic foundation of set theory.

" I. — L'intégrale indéfinie.

Soit $f(x)$ une fonction bornée intégrable définie dans (a,b); la fonction

$$F(x) = {_a}\!\int^x f(x)\,dx + K$$

est *l'intégrale indéfinie* de $f(x)$.

En appliquant le théorème de la moyenne on voit que *l'intégrale indéfinie de $f(x)$ est une fonction continue, à variation bornée, et qu'elle admet $f(x)$ pour dérivée en tous les points ou $f(x)$ est continue.*

Que se passe-t-il au point α si $f(x)$ n'y est pas continue? Alors il se peut qu'il y ait une dérivée égale à $f(\alpha)$; c'est le cas pour $\alpha = 0$ si $f(x)$ est nulle pour x quelconque, et égale à 1 quand x est l'inverse d'un entier; il se peut qu'il y ait une dérivée différente de $f(\alpha)$, c'est le cas pour $\alpha = 0$ quand $f(x)$ est partout nulle sauf pour $x = 0$; il se peut qu'il n'y ait pas de dérivée, c'est le cas pour $\alpha = 0$ quand $f(x) = \cos \mathfrak{L}\,|x|$ pour $x \neq 0$ et $f(0) = 0$.

Ainsi l'intégration peut conduire à des fonctions n'ayant pas partout une dérivée. Cette conséquence a été signalée par Riemann qui a appelé l'attention sur l'intégrale indéfinie de la fonction

$$f(x) = \sum \frac{(nx)}{n^2}.$$

Cette intégrale indéfinie $F(x)$ admet $f(x)$ pour dérivée quand x n'est pas de la forme $\dfrac{2p+1}{2n}$.

Supposons $\alpha = \dfrac{2p+1}{2n}$ et faisons tendre β vers α par valeurs croissantes, on a vu que $f(\beta)$ tend vers $f(\alpha) + \dfrac{\pi^2}{16\,n^2}$, donc, d'après le théorème de la moyenne, il en est aussi de même de $\dfrac{F(\beta) - F(\alpha)}{\beta - \alpha}$.

Au contraire, ce rapport tendra vers $f(\alpha) - \dfrac{\pi^2}{16\,n^2}$ si l'on fait tendre β vers α par valeurs décroissantes; donc $F(x)$ n'a pas de dérivée pour les valeurs de la forme $\dfrac{2p+1}{2n}$.

C'est le premier exemple que l'on ait connu d'une fonction de laquelle il n'aurait pas été clairement légitimé de dire qu'elle admet, *en général*, une dérivée. On connaissait bien des fonctions, celle de Cauchy, par exemple,

$+\sqrt{x^2}$, qui, en certains points, n'avaient pas de dérivée; mais ces points étaient exceptionnels, ils ne formaient jamais un ensemble partout dense; dans l'exemple de Riemann, au contraire, il y a des points sans dérivée dans tout intervalle. Le principe de condensation des singularités nous donnera autant d'exemples que nous le voudrons de fonctions analogues à celles de Riemann; si les a_p sont tous les nombres rationnels, $\int \sum \dfrac{\cos \mathfrak{L} \mid x - a_p \mid}{p^2} \, dx$ est une de ces fonctions."[31])

On the other hand, there exist functions which have a primitive but which are not Riemann-integrable. VOLTERRA gave the following example of such a function already in 1881 ([110]).

Let V be a non-dense perfect set in $[0,1] = I$, obtained in a well known way as the complement of a family of disjoint open intervals in I; let the measure of V be > 0 (see Hausdorff [57] p. 135).

Let α be any of these intervals. Consider the function φ

$$\varphi : x \to x^2 \sin \frac{1}{x} \quad (x \neq 0).$$

By means of φ we construct a function F on $\alpha = (a,b)$ as follows. One checks easily that the number

$$\xi = \max \left(x \in (a,b) \mid \varphi'(x-a) = 0, \ x \leqslant \frac{a+b}{2} \right)$$

exists. Put $\xi - a = t$.
We then define F on α by

$$F(x) = \begin{cases} \varphi(x-a) & \text{for } a < x \leqslant a+t \\ \varphi(t) & \text{for } a+t < x < b-t \\ \varphi(b-x) & \text{for } b-t \leqslant x < b. \end{cases}$$

Putting $F(x) = 0$ for $x \in V$, F is thus defined in I.
One shows easily that F is everywhere differentiable in I and that the derivative is bounded. F' is discontinuous in the points of V and is not Riemann-integrable.

---

[31]) $\cos \mathfrak{L} \mid x \mid$ means $\cos \log \mid x \mid$. Riemann designs by (x) the difference between x and the nearest integer and putting $(x) = 0$ if $x = n + \frac{1}{2}$. He defines the function $\Sigma \, \dfrac{(nx)}{n^2}$ in the same paper in which he defines his integral.

For functions of this kind a primitive cannot be calculated by means of Riemann integration.

Lebesgue studied in this respect also the function

$$f : x \rightarrow x^2 \sin \frac{1}{x^2} \quad (x \neq 0), f(0) = 0$$

and showed that the derivative is not $\mathcal{L}$-integrable (see Saks [98], p. 187).

He proved that with respect to the problem of the existence of primitive functions his integral has much better properties than the Riemann-integral. It would take too much place to give in detail the results of Lebesgue. He introduces a class of functions, now called *absolutely continuous functions*, a refinement of the notion of continuous function. Using functions of bounded variation, he comes to the so called *Lebesgue-decomposition theorem* of an interval function, leading to famous results about the problem of primitive functions.

I present some of them.

DEFINITION. *A function f is said to be absolutely continuous if to each $\varepsilon > 0$ there corresponds $\delta > 0$ such that*

$$\sum_{i=0}^{n-1} |f(a_{i+1}) - f(a_i)| < \varepsilon$$

for any finite sequence of non-overlapping intervals $[a_{i-1}, a_i]$ in $[a,b]$ satisfying

$$\sum_{i=0}^{n-1} |a_{i+1} - a_i| < \delta.$$

It is clear that any absolutely continuous function is continuous, and thus bounded, but the converse is not true. Any absolutely continuous function is a function of bounded variation (for the definition see below). Lebesgue showed that any function of bounded variation is almost everywhere differentiable, i.e., the points where it is not differentiable form a set of $\mathcal{L}$-measure 0.

Now, let F be an absolutely continuous function. Denote by $\dot{F}$ the function defined by

$$\dot{F}(x) = \begin{cases} F'(x) & \text{for any } x \text{ where } F'(x) \text{ exists,} \\ 0 & \text{everywhere else.} \end{cases}$$

One shows that the integral

$${}_a\!\int^b \dot{F}(x)\,dx$$

exists.

The following theorems can then be proved.

(1) *If F is absolutely continuous on* $[a,b]$ *then*

$$_a\!\int^x \dot F(t)\,\mathrm{d}t = F(x) - F(a)$$

for $a \leqslant x \leqslant b$.

(2) *Let f be a bounded function on* $[a,b]$; *suppose a function F exists such that* $F'(x) = f(x)$ *for all* $x \in [a,b]$. *Then*

$$_a\!\int^b f(x)\,\mathrm{d}x = F(b) - F(a).$$

The preceding example shows that these theorems are not valid for the Riemann-integral. However, one should want to have a theorem where – as in the second theorem – boundedness of f is not required. This was done by DENJOY and PERRON; I will speak about their theories later on.

For more information on the Lebesgue theory see Saks [98], Kamke [63].

Almost everywhere. The concept of a property valid almost everywhere (presque partout) was introduced by Lebesgue in the first edition of his "Leçons" (1903). In his thesis (1902) one finds only a restriction like "en faisant abstraction d'un ensemble de mesure nulle". Denjoy objected to the terminology "almost everywhere". There is a footnote about this in Lebesgue's Leçons (2e édition, p. 179) which I quote here because it is interesting with regard to the kind of the discussions:

"Nous conviendrons de dire qu'une propriété a lieu *presque partout* dans un intervalle (a,b), ou sur un ensemble $\mathscr{E}$, si les points de (a,b) ou de $\mathscr{E}$ en lesquels elle n'a pas lieu ou bien n'existent pas, ou bien forment un ensemble de mesure nulle.

Cette locution, introduite dans la première édition de ce Livre, a été généralement adoptée. Si l'on se rappelle que M. Denjoy n'a pas trouvé suffisamment précise et qu'il a rejeté l'expression: *le point P est point de l'ensemble E*, on ne s'étonnera pas qu'il ait jugé inadmissible la locution *presque partout* qui, à son avis, a deux sens: l'un qualitatif ou descriptif, l'autre quantitatif ou métrique. Je pense qu'il faut entendre par là qu'on aurait pu convenir de donner à *presque partout* la signification suivante: *exception faite des points formant un ensemble partout non dense.* Certes; mais M. Denjoy dit qu'une propriété a lieu sur une *epaisseur pleine* quand je dis qu'elle a lieu *presque partout. Epaisseur pleine* n'aurait-elle pas pu recevoir une autre signification que celle qu'il a plu à M. Denjoy de lui donner?

Presque partout serait inadmissible si, dans la langue usuelle, cette expression avait un sens précis, mais cela n'est pas; de sorte que le lecteur, en présence de l'énoncé précédent par exemple, ne peut lui donner aucun sens précis sans se reporter à la définition posée pour *presque partout*. Aucune erreur n'est donc possible.

Obliger le lecteur à se reporter à une définition a son inconvénient. Je l'accorderais volontiers à n'importe qui, sauf à M. Denjoy qui a utilisé dans ses Mémoires un nombre formidable de mots nouveaux. Et ce n'est pas diminuer l'inconvénient que de modifier, fût-ce même pour le perfectionner, un vocabulaire dont l'usage commence à se répandre".

As regards sets of measure zero, Borel (1898) states 'that "tout ensemble dénombrable a pour mesure zéro" and "tout ensemble dont la mesure n'est pas nulle n'est pas dénombrable. C'est surtout de cette dernière propriété que nous ferons usage".

Functions of bounded variation. The definition of the notion of a function of bounded variation is due to Jordan[32]).

DEFINITION. *Let f be a real function defined on the interval* $[a,b]$. *f is said to be of bounded variation if*

$$\sup \sum_{i=0}^{n-1} |f(a_{i+1}) - f(a_i)| < \infty$$

over all partitions $a_0 = a \leqslant a_1 \leqslant \ldots \leqslant a_n = b$.

It is easily shown that any function of bounded variation is the difference of two non-decreasing functions and, conversely, any such difference is a function of bounded variation. The theorem that any function of bounded variation is almost everywhere differentiable is much deeper; for a proof see Kamke [63]. Jordan introduced this concept in his study of the length of a curve, i.e. for the rectification of curves. He proved the following theorem:

If a curve C is given by the equations $x = \varphi(t)$, $y = \psi(t)$, $z = \chi(t)$, *then C is rectifiable on an interval* $[a,b]$ *if and only if* φ, ψ *and* χ *are of bounded variation.*

The length is defined as the limit of the lengths of polygons inscribed in *C*. Lebesgue continued this theory, finding the integral

$$_a\!\int^b ((\varphi')^2 + (\psi')^2 + (\chi')^2)^{\frac{1}{2}}\, dt$$

as an expression for the length, valid under more general conditions than the classical ones, which superfluously involved continuity.

Generalizations of functions of bounded variation are of importance in the further development of the theory of the integral (Denjoy).

---

[32]) See his Cours d'Analyse [62].

2.4   *More about Lebesgue*

For a good understanding of the development of analysis in the years round 1900, I must give more details about the work of Lebesgue. This is the more necessary because the work of Lebesgue was in the beginning accepted with reservation. I shall have to mention especially the different points of view of Borel and Lebesgue as concerns the foundations of analysis.

Of course I can give only short indications.

(i)   I mentioned the geometrical and the analytical definition of the Lebesgue-integral. It is interesting to see the way in which Lebesgue arrives at his analytical definitions. Lebesgue makes a distinction between *constructive* and *descriptive* definitions of mathematical notions.

A definition is called descriptive when the notion to be defined is characterized in it by means of certain properties which one wants to be satisfied.

A definition is called constructive when it indicates the operations to be performed in order to obtain the object.

Lebesgue remarks that, as to the descriptive method, the axioms to be posed must be consistent and, as to the constructive method, the operations must be possible.

The analytical definitions of the integral by Riemann and Lebesgue are constructive. The definition of primitive functions is an example of a descriptive definition.

Lebesgue gave a descriptive definition of an integral, which is interesting because it gives a first indication towards the modern definitions where an integral is defined as a mapping.

Lebesgue proposes to attach to every bounded function f on $[a,b]$ a real number, denoted by ${}_a\!\int^b f(x)\,dx$ and called the integral of f on $[a,b]$, in such a way that the following conditions are satisfied:

1.   For all a, b, h

$${}_a\!\int^b f(x)\,dx = {}_{a+h}\!\int^{b+h} f(x-h)\,dx;$$

2.   For all a, b, c

$${}_a\!\int^b f(x)\,dx + {}_b\!\int^c f(x)\,dx + {}_c\!\int^a f(x)\,dx = 0;$$

3.

$${}_a\!\int^b [f(x) + \varphi(x)]\,dx = {}_a\!\int^b f(x)\,dx + {}_a\!\int^b \varphi(x)\,dx;$$

4.   *If* $f \geqslant 0$ *and* $b > a$

$${}_a\!\int^b f(x)\,dx \geqslant 0;$$

111

5.

$$_0\!\int^1 1 \times \mathrm{d}x = 1;$$

6.   For any convergent sequence (f_n), tending to f, satisfying $f_1 \leqslant f_2 \leqslant \ldots$,

$$\lim_{n \to \infty} {}_a\!\int^b f_n(x)\,\mathrm{d}x = {}_a\!\int^b f(x)\,\mathrm{d}x.$$

As to the question of the independence of these axioms, Lebesgue remarks that it is easy to see that the first five conditions are independent (by constructing models satisfying four of them and not all five), but he does not know whether all six are independent (see his "Leçons").

   This problem was solved by BANACH [4][33]. He showed that it is possible to attach to *every* bounded function f and any interval $[a,b]$ a real number ${}_a\!\int^b f(x)\,\mathrm{d}x$ such that the conditions 1 – 5 are satisfied. If f is Riemann-integrable, this number is equal to the Riemann-integral of f, but for Lebesgue-integrable functions there is no such equality (the sixth condition is thus not satisfied). This proves that the condition 6 is independent of the conditions 1 – 5[34]).

   One should be inclined to discuss the definition of constructivity in the modern foundations of mathematics (intuitionism), but this should lead me too far from my subject. The work of Borel and especially the controversies between Borel and Lebesgue will allow me to say something about it later on.

   It is the aim of Lebesgue to give a constructive definition of the integral which is equivalent to the descriptive definition. This led him to the analytical definition mentioned before.

(ii)   There is a result of Lebesgue, which I must mention, though it is not in the domain

---

[33]   Lebesgue mentions this result in a note (p. 106) in the *second* edition of his "Leçons".

[34]   With this question we arrive in the domain of the foundations of mathematics. Tarski [106] proved that, given an infinite set E, there exists a non-trivial real measure, defined for *all* subsets of E, which is additive (for finite families) and which is 0 for any set consisting of one point and so for any finite set. Banach and Kuratowski [6] proved, accepting the continuum hypothesis, that this is not possible in general when the restricted additivity is replaced by the complete additivity (see my short note [81]).

The situation is still more complicated when it is required that, for instance in $\mathbf{R}^3$, congruent sets have equal measure. Compare the paradoxical decompositions of sets and spaces, mentioned in footnote 8.

The problem of the existence of sets which are not Lebesgue measurable is connected with these problems. Compare also the theory of the so called Banach limits, attaching a limit to *every* bounded sequence [5].

For results in this domain in connection with non-standard analysis see Luxemburg [79].

of the theory of the integral. It belongs to the theory of surfaces and I mention it here because of its relations with the discussions on functions which are not continuous.

A classical theorem in differential geometry states that a developable surface – that is the envelope of a one-parameter family of planes – can be mapped on a plane, that is can be put in one-one correspondence with a plane in such a way that any rectifiable curve on the surface is mapped on a rectifiable curve in the plane with the same length, and that, conversely, developable surfaces are the only surfaces which can be mapped on a plane in this way. This is shown in classical differential geometry under the assumption of continuity of the functions defining the surface. Analysing this theorem, Lebesgue observes that the restriction to continuous functions is not the right one for studying these isometric maps and he shows this by means of a surface, obtained by deformation of a scrap of paper, which evidently can be mapped isometrically on a plane but which is not a developable surface. This point of view differed essentially from the conventions of classical differential geometry. Burkill, in a necrology of Lebesgue [24], remarks that this note of Lebesgue seems to have "scandalisé Darboux".

It is beyond my subject to give the results which Lebesgue obtained in this domain; I refer the reader to his thesis. Let me just remark that Lebesgue showed by this example that the restriction to differentiable or even only continuous functions is not legitimate, but this was not generally accepted by the leading mathematicians.

(iii)   An other subject in this domain, studied by Lebesgue in his thesis, is the problem of defining the area of a surface. For plane surfaces it is the problem of measure, but for curved surfaces it involves serious difficulties.

In classical differential geometry the area of the surface S, defined by the equations $x = x(u,v)$, $y = y(u,v)$, $z = z(u,v)$ in some region R of the (u,v)-plane is, under wellknown assumptions on continuity and differentiability, proved to be equal to the integral

$$O(S) = \iint_R \left[\left(\frac{\partial(y,z)}{\partial(u,v)} \right)^2 + \left(\frac{\partial(z,x)}{\partial(u,v)} \right)^2 + \left(\frac{\partial(x,y)}{\partial(u,v)} \right)^2 \right]^{\frac{1}{2}} du \, dv.$$

By analogy with the definition of the length of a curve, it was long assumed that this value could be obtained as the limit of approximating polyhedrons, inscribed in the surface.

This idea was refuted by Schwarz, showing by an elementary example of a portion of a cilinder that the limit of inscribed polyhedrons may not exist and that it is even possible to choose a sequence of polyhedrons whose areas tend to any number not less than the actual area of the surface[35]).

---

[35])   Lebesgue remarks that Peano had already communicated this fact in his lessons in Turin (1881-1882) before the publication of the letter which Schwarz wrote about it to Genocchi. See [69].

Lebesgue studied the method of defining the area by means of approximating polyhedrons. He remarks:

"Une définition de l'aire qui ne s'applique qu'aux surfaces ayant des
tangents variant d'une façon continue, peut faire connaître une propriété géomé-
trique intéressante, mais ce n'est pas une véritable définition de l'aire, le nombre
à définir étant connu avant la définition."

Lebesgue formulates the problem in the following way:

"*Attacher à chaque surface un nombre positif fini ou infini que l'on
appellera son aire et satisfaisant aux conditions suivantes*:
1. *Il existe des surfaces planes ayant une aire finie.*
2. *Deux surfaces égales ont même aire.*
3. *Une surface somme de plusieurs autres a pour aire la somme des aires
   des surfaces composantes.*
4. *L'aire d'une surface S est la plus petite limite des aires des surfaces
   polyédrales dont S est la limite.*"

It is clear that a solution of this problem presupposes that a precise definition is given
of what is to be understood by "the surface S is the limit of polyhedrons P_n". This is
connected with the notion of the distance of two surfaces.

Lebesgue gives a solution, which turns out under appropriate assumptions to be the
classical integral. For details we refer the reader to the original paper of Lebesgue;
see Rado [93]; Saks [98]; for a modern treatment of these problems see Cesari [29a]
and Federer [41a].

Example (Schwarz-Rado).

Let S be the surface

$$x^2 + y^2 = 1, \quad 0 \leqslant z \leqslant 1.$$

Cutting S along the generator $x = 1$, $y = 0$, $0 \leqslant z \leqslant 1$ and mapping S on a plane, the image
of S is a rectangle A with sides 1 and 2π hence $O(S) = 2\pi$. Subdivide A in mn congruent
rectangles by subdividing the sides of A in m and n parts respectively. Subdivide any of
these rectangles in two triangles by drawing a diagonal; thus A is subdivied in $2mn$
triangles. On S the vertices of this network correspond to the vertices of a polyhedron
which is inscribed in S. The area of this polyhedron is calculated to be

$$O_{m,n} = 2n \ \sin \frac{\pi}{n},$$

tending to 2π for $m,n \to \infty$. Subdividing every rectangle in four triangles by drawing two diagonals, one obtains another inscribed polyhedron of which the area is calculated to be

$$O^*_{m,n} = 2n\ \sin\ \frac{\pi}{2n} + \left[\tfrac{1}{4} + \frac{4m}{n^4}\left(n\ \sin\ \frac{\pi}{2n}\right)^4\right]^{\tfrac{1}{2}} \cdot 2n\ \sin\ \frac{\pi}{n}.$$

If $m=n^3$ and $n \to \infty$, $O^*_{m,n}$ tends to ∞; if $m=n$ the limit is 2π. It is not difficult to see that any number a, $2\pi \leqslant a \leqslant \infty$, can be obtained as limit by choosing m and n in an appropriate way. Fréchet proved that in this example the cylinder may even be replaced by a polyhedron; see [46] for the papers of Fréchet on this subject.

2.5 *Controversies*

The work of Lebesgue was accepted by the leading mathematicians with a certain amount of mistrust. The reason was that discontinuous functions and functions without derivative were not considered as sound mathematical objects, leading to good mathematical theories. And precisely such functions play an important role in the work of Lebesgue. A qualification as "Mais une fonction a tout intérêt à avoir une dérivée", which I found in an article [38] of Montel on the work of Lebesgue and which, according to this author was due to Boussinesq, is a good illustration of the situation. Functions of this kind were considered as monstrosities and abnormalities and it was feared that the study of such objects should result in "une tératologie des fonctions[36]).

Some mathematicians studied discontinuous functions. I mention R. BAIRE (1874-1932; having studied at the Ecole normale supérieure he was professor in mathematical analysis in Dyon) who wrote a thesis on the subject "Sur les fonctions de variables réelles" (1900). But in his book "Leçons sur les fonctions discontinues" (1905) [3] he begins with an introduction in which he motivates the necessity of studying discontinuous functions, writing as follows:

"Si l'on jette un coup d'oeil sur le début d'un Cours d'Analyse classique, une chose ne manquera pas de frapper l'esprit. Les notions fondamentales sont présentées tout d'abord au moyen d'une définition extrèmement générale; puis, immédiatement après, des restrictions sont apportées à ces définitions, de manière à limiter le champ d'études, et c'est grâce à cette limitation qu'il est

---

[36] "Monstrosities" are also not unusual in topology. Compare the curious title of a paper of Dieudonné "Notes de Tératopologie (I)", Revue Rose), 1939, p. 39 (see Bourbaki [17], p. 25).

M A U R I C E   F R É C H E T

possible d'aller de l'avant et de construire les différentes théories qui constituent
la science mathématique.

Il est alors légitime de rechercher s'il n'est pas possible, en remontant aux définitions premières, d'en tirer des conséquences intéressantes tout en leur conser-
vant autant que possible leur généralité. On peut ainsi se proposer de constituer,
à côté de l'Analyse courante, une autre branche de l'Analyse, qui, bien entendu,
suivra de très loin la première, en tant que quantité de résultats acquis, mais qui,
en revanche, aura l'avantage de fournir des énoncés plus complets.

A cette partie des Mathématiques se rattachent les travaux, déjà nombreux,
faits en ces quarante dernières années, sur les fonctions discontinues, les
fonctions sans dérivées, les fonctions pourvues de dérivées de tous ordres,
mais non développable en série de Taylor, l'intégration des fonctions les
plus générales, la définition générale des courbes fermées dans le plan, etc.
Il est bien remarquable d'ailleurs que l'Analyse courante ne peut pas indéfini-
se passer des considérations qui font l'objet de la branche dont nous parlons.
Les singularités de toutes sortes, les discontinuités, par exemple, s'introduisent
d'elles mêmes, qu'on le veuille ou non, dans des questions d'où le chercheur
aurait souhaité les écarter.''

Baire even gives motivations from the side of physics and mechanics.

This mistrust has lasted for years. I found the following curious passage in a book
of P. Boutroux, ''L'idéal scientifique des mathématiciens dans l'Antiquité et dans les
temps modernes'' ([20], 1920).

He observes that the fact that in mathematics one theory is sometimes found to be
more interesting than an other is mostly a matter of prejudices. He continues (l.c. p. 260):

''M. Denjoy le montre une fois de plus, fort spirituellement, dans un récent article (1)[37]), ou il raille les savants traditionalistes de ne s'intéresser qu'à de "bonnes bourgeoises de fonctions" et de méconnaître l'importance de certains travaux récents. Ces esprits retardataires semblent en effet se méprendre complètement sur la mission qui incombe au véritable analyste. La découverte doit être, selon nos vues, une exploration; le mathématicien a pour mission de rechercher ce qui *est*; son but est de dresser la carte du monde des faits mathématiques."

Now, Denjoy has done important work on the theory of integration after Lebesgue, so it should be possible that the words of Denjoy concern those mathematicians who did not fully accept the work of Lebesgue (compare later on the controversy between Borel and Lebesgue).

The integral of Lebesgue has only very gradually taken the place of the Riemann-integral, in spite of its better properties. Even in the twenties, the Riemann-integral dominated and generally the Lebesgue-integral was not taught at the universities. As to the opinion of Lebesgue himself, the following indication, which I found in [38], p. 16, is noteworthy:

"Un jour que de jeunes mathématiciens réunis chez moi (i.e. Montel) discutaient sur l'opportunité d'enseigner dès le début l'intégrale de Lebesgue, celui-ci arriva inopinément. On lui posa la question: "Par quelle intégrale doit-on commencer devant de jeunes étudiants? – Par celle de Riemann, bien entendu" répliqua Lebesgue."[38])

A further question I have to speak about is the controversy between Borel and Lebesgue.

On the one hand there is the question of priority between Borel and Lebesgue as concerns the definition of the integral. On the other hand there was a controversy on questions concerning definition of mathematical objects, I mean questions on constructive definitions, descriptive definitions and definitions permitting effective calculation. They are in the domain of the foundations of mathematics. In the controversy between Borel and Lebesgue these two questions (priority and definitions) are interwoven and can scarcely be separated.

---

[37]) A. Denjoy. L'Orientation actuelle des Mathématiques, Revue du Mois, 10 avril 1919. I found it impossible to obtain this article. This journal was founded in 1906 by Emile Borel. After the war 1914-1918 Borel ceased the publication owing to financial difficulties; see [47].

[38]) Lebesgue was interested in questions concerning secondary education. For many years he was a member of the editorial staff of the journal "L'Enseignement mathématique".
In his later years he published on the domain of elementary geometry; we mentioned already his work on the problem of the equivalence of polyhedrons ([73]).

In 1912 Borel [13] wrote a large paper in which he states his point of view in these fundamental questions.

It is Borel's opinion that only numbers and functions which can effectively be calculated are useful (apart perhaps from theoretical considerations). A number α is called *"calculable"* if "étant donné un nombre entier quelconque *n*, on sait obtenir un nombre rationnel qui diffère de α de moins de $\frac{1}{n}$". A function f is calculable if its value is calculable for any value of the variable which is calculable. "Une fonction ne peut donc être calculable que si elle est continue, au moins pour les valeurs calculables de la variable".

He makes several remarks concerning definitions which can be allowed in mathematics and those which can not. Borel does not accept a definition like: "the number *a* is equal to 0 if the constant of Euler *C* is an algebraic number and equal to 1 if not".

Note that it is an open problem whether the constant of Euler, which is defined by

$$C = \lim_{n \to \infty} \left(\sum_{k=1}^{n} \frac{1}{k} - \log n \right),$$

is an algebraic number or not[39]).

On the base of this opinion Borel gave a new definition of the integral, using approximation by polynomials. He was of the opinion that this method was better than that of Lebesgue.

Lebesgue answered Borel in an extensive paper [72] in 1918, in which he explains that Borel is mistaken. He comments in this article also on the question of priority, which Borel seemed to claim. Borel, however, had published nothing on the integral before Lebesgue. Lebesgue discussed questions such as: what is primary: measure or integral, that is to say who was the first to observe that measure is a special case of integration by considering characteristic functions?
Borel replied in a paper of 1919.

We must content ourselves with these short indications; those who are interested should read the original papers.

2.6 *Real functions*

Although it is not strictly in the domain of the history of the integral, it is inevitable

---

[39]) The reader recognizes in these short indications the intuitionistic point of view. Borel, Lebesgue, Baire and Lusin are known as semi-intuitionists. For more information see Bockstaele [8], which contains an extensive bibliography.

VITO VOLTERRA

to make some remarks on the development of the analysis, in particular the theory of real functions, in the nineteenth and the beginning of the twentieth century. It will be clear from the preceding sections that both subjects can hardly be separated.

It is interesting to read what Boutroux says in the book which I already mentioned [20] about this period. He states that the important place of algebra is diminishing in the nineteenth century. In that century the study of functions takes a dominating place. He mentions (p. 171) the following assertion of VITO VOLTERRA, a great Italian mathematician who has had much influence on the development of analysis: "Je n'ai pas hésité en 1900, au Congrès des Mathématiciens de Paris, à appeler le 19e siècle le siècle de la théorie des fonctions" (apud Henri Poincaré, Alcan, 1914, p. 14); see also [19].

It suffices to see the articles in the leading mathematical journals round 1900 to ascertain the important place of the study of real functions. The theory is of course connected with the development of set-theory by Cantor. I mention for instance the problem of the classification of real functions for which transfinite numbers are important. The question of the meaning of what a good function really is, was still a point of discussion and continuous nowhere differentiable functions were considered as more or less pathological objects. Borel, for instance, writes as follows in his 1912-article ([13])[40]:

---

[40]   Borel added these remarks as Note VI to the second edition (1914) of his book Leçons sur la théorie des fonctions (first edition 1898); the article is then called: "La théorie de la measure et la théorie de l'intégration".

119

"Les résultats acquis, dès la fin du XIXe siècle, ont surabondamment prouvé combien était simpliste l'opinion d'après laquelle il serait possible de limiter le champs des Mathématiques à l'étude d'une catégorie déterminée de fonctions: fonctions continues, fonctions dérivables, fonctions analytiques, etc. ...."([13]. p. 159).

But, having given an example of a differential equation which has no analytical solution, he continues (l.c. p. 160):

"Dans un autre ordre d'idées, M. Lebesgue a tiré de la considération des développements décimaux des nombres irrationnels les plus généraux, des conséquences presque paradoxales, en particulier il en a déduit la définition d'une fonction qui n'est susceptible d'aucune représentation analytique."

In a footnote he remarks, however, that these results must be judged with reserve and that in his opinion the interest of such results is mostly negative.

Borel continues as follows (l.c. p. 160):

"C'est cette quasi-impossibilité d'établir une démarcation précise entre les êtres analytiques regardés comme "simple" et les autres, qui a été l'origine de travaux qui ont considérablement accru nos connaissances en Analyse. Ces travaux étaient nécessaires; ils ne sont pas d'ailleurs définitifs sur tous les points et il sera encore utile, à mon avis, de s'occuper de ce qu'on a pu appeler la *pathologie* des fonctions. Mais il est permis de penser que le but définitif de ces recherches *pathologiques* doit être la délimitation des fonctions considérées comme *saines*. Là encore, nous nous heurtons à des difficultés qui sont loin d'être résolues."

I mention the titles of some other notes of Borel which will give an impression of the kind of subjects which are treated: La notion des puissances, la puissance des ensembles de fonctions. La croissance des fonctions et les nombres de la deuxième classe. La notion de fonction en général; les fonctions discontinues; les fonctions définies par des conditions dénombrables; la notion de fonction arbitraire[41]).

Sur les définitions analytiques et sur l'illusion du transfini. Sur la classification des ensembles de mesure nulle.

Sur l'intégration des fonctions non-bornées et sur les définitions constructives.

Transfinite numbers and transfinite induction are frequently used in these articles.

---

[41]) Even in the thirties one finds notes about the meaning of an arbitrary function. See for instance K. Knopp, Theorie und Anwendung der Unendlichen Reihen (Berlin 1922), footnote p. 339.

Nowadays, Zorn's lemma – which was not yet known in this period – is generally used instead of transfinite induction.

Transfinite induction.   Transfinite induction is a generalisation of the common method of induction to propositions depending on ordinals. Baire [3] states it as follows. (p. 47):

"On sait quelle est, en mathématiques, l'importance du procédé de démonstration dit de récurrence, qui permet d'affirmer qu'une proposition est vraie pour un nombre *n* susceptible de prendre toutes les valeurs entières positives, si l'on démontre: 1. que la proposition est vraie pour les premières valeurs de *n*; 2. que, si elle est vraie pour tous les entiers qui précèdent un nombre *n*, elle est encore vraie pour *n*. Nous allons, en ce qui concerne les nombres des classes I ou II, indiquer un procédé analogue.

THÉORÈME.   *Si, dans l'énoncé d'une proposition, figure un nombre α susceptible de prendre toutes les valeurs des nombres des classes I et II, cette proposition sera démontrée pour toutes ces valeurs, si l'on montre: 1. que la proposition est vraie pour les premières valeurs de α (pour α = 0 ou pour α = 1); 2. que, si la proposition est démontrée pour tous les nombres α' inférieurs à un nombre déterminé α, α étant un nombre quelconque des classes I ou II, elle est encore vraie pour la nombre α.*"

In modern notation this is:

$$(\forall \beta < \alpha \; A(\beta) \to A(\alpha)) \to \forall \alpha \; A(\alpha).$$

There are many discussions in the litterature of that time on the validity of the principle of transfinite induction, which belong to the domain of the foundations of mathematics *Zorn's lemma* is the following statement:

If every subset of a partially ordered set A which is linearly ordered in the induced order has an upper bound in A, then A contains a maximal element, that is an element x such that if y∈A and x ≤ y, then y = x.

I don't treat the connections between this lemma, the Axiom of Choice and the well-ordering-theorem.

The years round 1900 mark the beginning of a series of books in this domain of analysis called "Collection de monographies sur la théorie des fonctions", also known as the "Collection-Borel" because Borel took the initiative in publishing these series. Borel himself wrote several books in this series. I mentioned already his "Leçons sur la théorie des fonctions" (1898 and 1914)[42]).

---

[42])   Fréchet [47], p. 17, remarks that Borel took the initiative for publishing this "Collection" in 1904.

Further there are Borel's books:

Leçons sur les fonctions entières (1900).

Leçons sur les séries divergentes (1901).

Leçons sur les fonctions de variables réelles et les développements en séries de polynomes (1905).

Méthodes et problèmes de théorie des fonctions (1922).

The famous book of Lebesgue (Leçons [70]) is in this series; also Baire [3].

With regard to integration theory I mention the treatise of De La Vallée Poussin [108].

Several of these books contain introductions to set theory and chapters on the representation of continuous and discontinuous functions by means of series of polynomials, to be seen in the light of the old discussions about the concept of a function (compare the classical theorem of Weierstrass on the approximation of continuous functions by means of polynomials).

There are many historical notes in these books and articles which are interesting to read for those who are interested in the historical development of analysis. They are still worthwhile to read, be it only to become aware of the fact that there is continuity in the development of mathematics.

This was also the period of the famous *Cours d'Analyse* and *Traité d'Analyse*, which covered in big volumes the various domains of analysis. I mentioned already CAMILLE JORDAN; further there are HERMITE, GOURSAT, PICARD, DE LA VALLÉE POUSSIN.

Classification of Baire (classes de Baire). By way of example I treat the classification of Baire of discontinuous functions. I mentioned this mathematician before and his defense of the use of discontinuous functions. He gave the following classification.

The *class* 0 contains all continuous functions.

The *class* 1 contains the limits of all sequences of functions of class 0, provided these limits don't belong to the class 0.

It is obvious how, for any integer n, the *class of ordinal n* is to be defined.

Having thus defined the classes for all finite ordinals, the *class of order ω* is defined as the set of all functions which are the limit of a sequence (f_n) of functions belonging to classes of finite order, provided this limit function does not belong to any class of finite order.

In a well known way, this method can be continued transfinitely, thus obtaining classes for any enumerable ordinal α, finite or transfinite.

However Borel's book ,,Leçons sur les séries divergentes" from 1901 was already written under the head Collection de monographies sur la théorie des fonctions publiée sous la direction de M. Emile Borel.

The sets obtained this way are the so called *classes of Baire*. The functions are known as *"fonctions de Baire"*[43]).

The problem then is: *given an enumerable ordinal α, do there exist functions of class α?*

This problem was treated by Baire, Borel, Lebesgue and De La Vallée Poussin and the answer is: yes, for any α there are functions of class α.

As Borel remarks [12], this is easily seen by an argument on the cardinal numbers of the classes. However, he did not accept such a reasoning as a rigorous proof; in his opinion it is necessary to define effectively a function for any class. The ideas of Borel are in Note III (Sur l'existence des fonctions de classe quelconque) of his book on the theory of functions [12], from which I take the following part. (l. c. p. 156).

"On peut se demander si la classification de M. Baire n'est pas purement idéale, c'est-à-dire *s'il existe* effectivement des fonctions dans les diverses classes définies par M. Baire. Il est clair, en effet, que si l'on prouvait, par exemple, que toutes les fonctions sont de classe 0, 1, 2 ou 3, la plus grande partie de la classification de M. Baire serait sans intérèt. Nous allons voir qu'il n'en est rien; mais il est tout d'abord nécessaire d'insister un peu sur ce que l'on doit appeler une fonction *définie*. Il est, en effet, aisé de voir qu'il *existe* des fonctions de classe supérieure à un nombre quelconque (fini ou transfini) donné d'avance; car l'ensemble *E* des fonctions dont la classe ne dépasse pas un nombre donné a évidemment la puissance du continu; et l'ensemble *F* de toutes les fonctions possibles a une puissance supérieure à celle du continu; l'ensemble *F* est donc infiniment plus riche que l'ensemble *E*, c'est-à-dire renferme une infinité de fonctions qui n'appartiennent pas à *E*. Mais ce raisonnement basé sur les puissances a un grave défaut: il nous apprend bien qu'il y a des fonctions de *F* qui n'appartiennent pas à *E*, mais il ne nous donne pas le moyen *d'en définir une*, c'est-à-dire d'en désigner *une* de telle manière qu'on puisse la distinguer des autres; en d'autres termes de manière que deux personnes différentes, lorsqu'elles parlent de cette fonction, soient certaines qu'elles parlent de *la même*.

Le raisonnement précédent ne permet pas d'exclure l'hypothèse ou un théorème tel que le suivant serait exact: *toute fonction effectivement définie est nécessairement de classe* 0, 1, 2 *ou* 3. Nous allons, au contraire, montrer qu'il est possible de *définir* effectivement une fonction dont la classe dépasse un nombre donné d'avance."

---

[43]) These are the functions which Lebesgue calls "fonctions représentables analytiquement". He defines them as functions which can be obtained by the repeated application of certain operations (addition, taking limits, etc.), starting from elementary functions. See his article in ,,le journal de M. Jordan" [71].

It would take too much place to give the theory of Baire-functions; the examples are rather complicated. The Dirichlet-function, which I defined in 1.3 belongs to class 2.

Lebesgue [71] showed that there exist functions which don't belong to any class; such a function "n'est susceptible d'aucune représentation analytique" (compare the notes on Borel in 2.6). He also proved that the Baire-functions are identical with the B-measurable functions and thus, the problem of the existence of functions which belong to no finite or infinite class is equivalent to the problem of the existence of sets which are not B-measurable. Lebesgue (1905) gave an example of a measurable set which is not B-measurable; but he remarks: "Mais la question beaucoup plus intéressante: peut-on nommer un ensemble non-mesurable? reste entière". We already mentioned this problem and the solution.

I note that any bounded Baire-function is Lebesgue-integrable; this follows from the Lebesgue-convergence-theorem.

For an excellent exposition of the theory of these functions see De La Vallée Poussin [108]. For generalisations see [41a].

Of course I have not been complete in reviewing the names of the mathematicians who in the first decades of our century did fundamental work in the domain of set theory, the theory of real functions, integration theory and in a domain which I did not yet mention but which is connected with our subject, namely general topology. But I cannot omit mentioning the famous Polish school with names like Sierpinski, Lusin (Russian), Souslin, Banach, Steinhaus, Mazurkiewicz, Kuratowski and others. Much of their important work is published in the journals Fundamenta Mathematica and Studia Mathematica.

But I return now to the integral.

2.7   Denjoy

I return to the problem of primitive functions. When this problem is put in the following way:

"*Given a function φ, find a function f which has φ as derivative*", there is evidently no solution in general, because a function which is a derivative is at most of class 1.

But even when it is known that a function φ is the derivative of another function f – this necessitates the characterization of such functions–the increment of f, i.e., $f(b)-f(a)$, cannot always be calculated by means of an integration, because φ may not be integrable. We have given an example in 2.3 (iii).

We saw that the Lebesgue-interval provides better results than the Riemann-integral, but even Lebesgue has to put restrictions on the derivative to be integrated (boundedness, absolute continuity, see 2.3 (iv)). Thus, the Lebesgue-integral does not provide

ARNAUD DENJOY

us with an integral as the inverse operation of differentiation for the most general case.

It is DENJOY who gave the solution of the problem. The method of Denjoy consists in defining a finite or transfinite hierarchy of continuous operations (integrations), starting with the Lebesgue-integral and passing from one stage to another by certain operations, by which it is possible to determine the primitive functions of any function which is a derivative. He called this sequence "*totalisation*". When the operations are applicable to a function f, f is called *totalisable* and the result is a function F of x, called the "*total indefini*" of f; it is determined up to a constant. The increment of F in an interval (a,b) is called the *total* of f. Observe the agreement between the idea of this *constructive* definition of an integral and the idea of the Baire-hierarchy of real functions. See Denjoy [34], [35], [36]; also Lebesgue "Leçons" 2nd edition).

The integral, defined in this way, is nowadays mostly called the *Denjoy-integral*. For a descriptive definition of this integral see for instance Saks [98]. Generalizations of functions of bounded variation and absolutely continuous functions are used. It is not my intention to give here the definitions; I only mention that *Denjoy-integrals in the wide sense* (Denjoy-Khintchine) and *Denjoy-integrals in the restricted sense* (Denjoy-Perron) are defined.

Reading in the original papers of Denjoy, one finds the expressions "somme bes-gienne", "intégrale besgienne"; the reader understands of course what Denjoy means by them. Montel [38] writes about it as follows.

"M. Denjoy avait, après Lebesgue, donné une puissance nouvelle et définitive à la notion d'intégrale par sa belle création de la totalisation. Dans

Oskar Perron

ses travaux, il introduisit le qualicatif "besgien" pour désigner l'intégrale de Lebesgue qui devient l'intégrale "besgienne". Lebesgue, mécontent, lui écrivit une de ces longues lettres dont il était prodigue, humoristique, ironique et mordante. "Vous appelez mon intégrale besgienne, écrivait-il, que diriez-vous si j'appelais la vôtre joyeuse?" "Vous voulez appeler mon intégrale joyeuse, répondit M. Denjoy, je vous en défie bien"."[44])

For information on the work of Denjoy – who was a professor at the University of Utrecht from 1917 to 1922 and afterwards a professor at the Faculté des Sciences de Paris – see [37].

2.8 *Perron*

There is an other way of treating the problem of primitive functions which I must mention because it is entirely different from the previous methods.

The methods, hitherto described, all require measure theory. PERRON [91] gave a

---

[44] There are more interesting personal remarks in the paper [38].

method of defining an integral not requiring measure. The principle is again that of defining an object by means of approximation from below and from above, now with special regard on the derivatives of a function. Without entering into details, his method is the following (see Kamke [63]).

I consider the extended real line $\bar{\mathbf{R}}$, that is I introduce two new elements $+\infty$ and $-\infty$ with the usual conventions:

$$a + \infty = +\infty + a = +\infty, \ +\infty + \infty = +\infty,$$
$$a - \infty = -\infty + a = -\infty, \ -\infty - \infty = -\infty,$$
$$+(+\infty) = -(-\infty) = +\infty, \ +(-\infty) = -(+\infty) = -\infty, \ -\infty < a < +\infty.$$

Let φ be a finite function defined on the interval $[a,b]$.

DEFINITION. *The lower derivative of φ in $\xi \in [a,b]$ is defined by*

$$\underline{D}\varphi(\xi) = \lim_{x \to \xi} \inf \frac{\varphi(x) - \varphi(\xi)}{x - \xi}.$$

The upper derivative is defined by

$$\bar{D}\varphi(\xi) = \lim_{x \to \xi} \sup \frac{\varphi(x) - \varphi(\xi)}{x - \xi}.$$

They may be $+\infty$ or $-\infty$.
The following properties are evident:

$$\underline{D}\varphi(x) \leqslant \bar{D}\varphi(x),$$
$$\underline{D}(-\varphi(x)) = -\bar{D}\varphi(x),$$
$$\bar{D}(\varphi(x) + \psi(x)) \leqslant \bar{D}\varphi(x) + \bar{D}\psi(x), \bar{D}\varphi(x) \neq -\infty, , \bar{D}\psi(x) \neq -\infty,$$
$$\underline{D}(\varphi(x) + \psi(x)) \geqslant \underline{D}(x) + \underline{D}\psi(x), \underline{D}\varphi(x) \neq +\infty, \underline{D}\psi(x) \neq +\infty.$$

DEFINITION. *Let f be a function aefined on the interval $[a,b]$. A function φ, defined on $[a,b]$, is called a minor function of f if*

(i)   *φ is finite and continuous on $[a,b]$,*
(ii)   *$\varphi(a) = 0$,*
(iii)   *$\bar{D}\varphi(x) \neq +\infty$ for all $a \leqslant x \leqslant b$*
(iv)   *$\bar{D}\varphi(x) \leqslant f(x)$ for all $a \leqslant x \leqslant b$.*

A function ψ, defined on $[a,b]$ is called a major function of f if

(i)   *ψ is finite and continuous on $[a,b]$,*
(ii)   *$\psi(a) = 0$,*
(iii)   *$\underline{D}\psi(x) \neq -\infty$ for all $a \leqslant x \leqslant b$,*
(iv)   *$\underline{D}\psi(x) \geqslant f(x)$ for all $a \leqslant x \leqslant b$.*

A. F. Monna

THEOREM. *Let φ be a minor function and ψ a major function of f. Then*

(i) $\varphi(x) \leqslant \psi(x)$, $a \leqslant x \leqslant b$,

(ii) $\psi - \varphi$ *is a non-decreasing function.*

Not every function has minor and major functions[45]).

Example. The function f, defined on $[0,1]$ by $f(x) = \dfrac{1}{1-x}$ for $x \neq 0$ and $f(1) = 0$ has no major functions.

Now, let f be a function having minor and major functions φ and ψ respectively.

DEFINITION. *Put*

$$\Phi(x) = \sup_{\varphi} \varphi(x).$$

$\Phi(b)$ *is called the lower Perron integral of f on $[a,b]$.*

Similarly, if

$$\Psi(x) = \inf_{\psi} \psi(x),$$

then $\Psi(b)$ is called the upper Perron integral of f.

Notation:

$$\Phi(b) = (P)_a\underline{\int}^b f(x)\mathrm{d}x,$$

$$\Psi(b) = (P)_a\overline{\int}^b f(x)\mathrm{d}x.$$

It is evident that

$$(P)\underline{\int} f(x)\mathrm{d}x \leqslant (P)\overline{\int} f(x)\mathrm{d}x.$$

The function f is said to be *integrable in the sense of Perron* if (i) f has minor and major functions and (ii) the lower and the upper Perron integral of f are equal. This common value is called the definite Perron-integral of f over $[a,b]$ and is denoted by

$$(P)_a\int^b f(x)\mathrm{d}x.$$

Because a function f, which is P-integrable over $[a,b]$, is also P-integrable over any interval contained in $[a,b]$, the definition of the indefinite Perron-integral is evident.

---

[45]) Minor and major functions were first introduced by De La Vallée Poussin in his Cours d'Analyse.

I mention the following properties of the Perron-integral:

(i)   *If the function f is integrable in the sense of Lebesgue over* $[a,b]$ *then it is integrable in the sense of Perron and the integrals are equal. For bounded functions the converse is true.*

For bounded functions the Lebesgue- and the Perron-integral are therefore equivalent. But the Perron-integral includes more: one can prove that it includes also the improper Riemann-integrals, which fails for the Lebesgue-integral.

(ii)   *Suppose F has a finite derivative for any* $x \in [a,b]$:

$f(x) = F'(x), x \in [a,b]$.

Then f is Perron-integrable and

$(P)\ _a\!\int^b f(x)\,dx = F(b) - F(a)$.

Note that it not required now that f is bounded.

It follows from these properties that the Perron-integral gives a synthesis of the classical method (Newton) and the method of approximating sums (Lebesgue).

The connection between the Denjoy-integral and the Perron-integral was studied by several mathematicians. The integration in the sense of Perron is equivalent with the integration in the restricted sense of Denjoy. I refer the reader to Saks [98]. It is clear that these theories are in close connection with the theory of real functions.

Remark.   Perron used an analogous method in potential theory for his study of the Dirichlet problem. Using subharmonic functions (minor functions) and superharmonic functions (major functions), he showed with an analogous approximation method the existence of a certain harmonic function, the so called generalized solution of the Dirichlet-problem.

2.9   *Further developments*

I will make some supplementary remarks on the developments in integration theory preceding the modern theory.

First I have to mention that G. VITALI (1904) and W. H. YOUNG (1904) independently from Lebesgue and from each other came to a definition of measure analogous to the measure of Lebesgue (for literature see Saks [98]).

The problems which were studied in connection with the Lebesgue-integral are well known: sets of measure 0, the notion almost everywhere, functions which are almost everywhere defined leading to the normed linear space L^1 of the Lebesgue integrable functions and to the normed linear space L^p ($1 \leqslant p < \infty$) of the measurable functions f for which $_0\int^1 |f|^p dx < \infty$. I will not go into these matters; they can better be treated in a history of functional analysis, which is in close relation with the development of the integral.

The theory of functions of a set – in particular functions of an interval – was a subject of research. If f is a function which is integrable on $\mathbf{R}^n$, or on a convenient open set in $\mathbf{R}^n$, then the integral

$$F(E) = {}_E\int f(x)\,dx$$

is an additive set function which is defined for measurable sets E. But set functions can be defined in an abstract way, without using an integral; complete additivity is postulated. Points of research are continuity, representation of set functions by an integral. Also the problem of differentiability, where, given a point $x \in \mathbf{R}^n$, one studies the existence of the limit

$$\lim_{E \to x} \frac{F(E)}{m(E)}$$

where $m(E)$ is the measure of E and E tends to x in an appropriate way; this is the differentiation in measure.

This subject belongs to the general theory of measure, nowadays fundamental for the modern theory of probability. Here measure is a subject of research for itself; note that for Lebesgue measure was only a notion which he needed to arrive at the definition of the integral.

In the further development I have to mention explicitly the names of some mathematicians who contributed in an important way to the theory.

First I mention THOMAS JAN STIELTJES (1856-1894). Born in Zwolle in the Netherlands, Stieltjes went to Paris in 1885 after a somewhat troublesome start – he studied engineering without success, worked for some time in the astronomical observatory in Leiden but turned to mathematics; he obtained the doctors degree in France in 1886 on a thesis "Séries semi-convergentes". In that year he went to Toulouse, where he became professor in mathematics at the university.

In relation with his research in quite another domain, the theory of continued fractions and the problem of moments (for a review see [82]), Stieltjes came to define a new notion of integral, now known under the name *Stieltjes-integral*. He published his results in a famous paper in 1894 [102] "Recherches sur les fractions continues".

Stieltjes observed that the method of Riemann for defining the integral of a continuous function can be performed in exactly the same way when, instead of using the ordinary measure on the real line, the increments of a non-decreasing function are used as a

THOMAS JAN STIELTJES

measure for the real line. Introducing the sums, just as for the Riemann-integral,

$$\sum_{i=1}^{n} m_i(g(x_i) - g(x_{i-1})),$$

$$\sum_{i=1}^{n} M_i(g(x_i) - g(x_{i-1})),$$

he obtains thus an integral for a continuous function f in the well known notation

$$_a\int^b f(x)\,dg(x).$$

This is in an evident way generalised for the case of a function g of bounded variation, such a function being the difference of two non-decreasing functions.

The essential fact is that in this way instead of one integral, as in the classical theory, a family of integrals is assigned to a continuous function f. This notion includes the Riemann-integral for continuous functions.

I hardly need to indicate the physical interpretation of such an integral: if g is a non-decreasing function, the integration of f is performed with respect to a distribution of positive mass, where the points of discontinuity of g correspond with the finite masses concentrated in a point [46]). The Stieltjes-integral is of importance in modern potential theory; I mention for instance the representation of harmonic functions in the form of an integral (compare the Poisson-integral).

With his work in this domain Stieltjes is one of the mathematicians who opened the way to the modern functional analytic approach to the integral.

Stieltjes was a most remarkable mathematician. Famous, perhaps unique, is his correspondence with HERMITE. From 1882 to 1894 they exchanged 432 letters, all with mathematical content (analysis). These letters are published [1]; in the first volume one finds a biography of Stieltjes.

In 1913 RADON defined an integral in $\mathbf{R}^n$, not starting, as Lebesgue does, with the Lebesgue-measure, but with a completely additive set function [94]. This integral includes the Lebesgue-integral as well as the Stieltjes-integral (for continuous functions) and it thus provides a synthesis of the older theories. This leads to the *Lebesgue-Stieltjes-integral*, no longer defined only for continuous functions but for a larger class of functions.

We see here two aspects of the theory of the integral: on the one hand the question of the extension of an integral to a class of functions which is as large as possible, and on the other hand the extension of the concept of integral.

Hitherto the definition of the integral was restricted to functions defined on $\mathbf{R}^n$. FRÉCHET remarked that this restriction is not necessary and that the integral of Radon can be defined for real functions defined on an abstract set, even without the necessity of defining a topology on this set (see [44]). This was an important remark because it opened the way to further generalizations leading, for instance, into the domain of an abstract kind of analysis.

Fréchet is one of the pioneers in the domain of what is called *"analyse générale"*. He observed in 1904 that for some theorems in classical analysis it is not necessary that the variables are real numbers. They may be elements of an abstract set, provided where necessary with a certain structure, for instance a topology or an algebraic structure. This leads to *general analysis*, from which real analysis is a special case. Fréchet published many papers on this new domain of mathematics. He introduced and studied, for instance, in 1906 the general notion of distance on an abstract set, leading to the study of metric spaces, a very important notion in modern mathematics. Fréchet

---

[46] Before Stieltjes Cauchy already had the idea of integration with respect to a function. See his paper "Sur le rapport différentiel de deux grandeurs qui varient simultanément"; Oeuvres, 2e série, t. XII, 214-262. For information on this work of Cauchy, see Lebesgue "Leçons" (2e ed.) p. 290.

JOHANN RADON

published also on several other domains of mathematics (for instance on probability). We refer the reader to [43], [45], [46].

Independently from Fréchet, E. H. MOORE published in 1906 results on general analysis [85].

All these were steps towards the definition of an integral as a mapping which I am going to treat now.

3 Modern theory of the integral

3.1 *Integrals as linear functionals*

The theory of the integral, which I sketched above, had an essentially geometrical charac-ter, that is, it was based on the notion of area, that is to say, on measure (in the theory of the Perron- and the Denjoy-integral the problem of primitive functions is placed in the foreground).

As I said before, the theory of measure and the theory of integration founded on measure has continued its development, but then in axiomatic form. I don't treat this subject in this book, because it should be chapter for itself (there is however in one of the following paragraphs some information about Carathéodory). In modern times there is an other direction in the theory of the integral: *it is that direction in which the notion of mapping is central.*

In elementary mathematics, for instance, one can go the way of introducing the notion of volume as a mapping of polyhedrons in the set of positive real numbers which has certain properties, such as additivity for disjoint polyhedrons.

This new course in the theory of the integral cannot be isolated from the development of abstract analysis in the twentieth century. I already mentioned Fréchet as a pioneer in this domain. I may refer to the theory of Hilbert-space, to the Banach-spaces and the locally convex spaces (which need not even be metric spaces), the theory of linear mappings, the theory of duality and so on. The modern theory of the integral which I have in view must be placed in this frame.

In section 2.6., I quoted Volterra, who characterized in 1900 the nineteenth century as the century of the theory of functions; he observed that in that century the influence of algebra was diminishing. It is interesting to mention here the famous book of LA-GRANGE, published in 1797[47]) "Théorie des fonctions analytiques, contenant les principes du calcul différentiel, dégagés de toute considération d'infiniments petits ou d'évanouissans, de limites ou de fluxions, et réduits à l'analyse algébrique des quantités finies". He reduced differential calculus to algebra, defining differentiation in a formal way by considering the power series for an analytic function.

It is perhaps too early to characterize the first half of the twentieth century, but I am inclined to say that there is a strong tendency towards the application of algebraic methods in analysis. There are plenty of examples; it suffices to look in modern books on analysis to perceive this tendency. E.g. I mention linear analysis, where the linear structure of certain sets of functions is an object of study, for instance the spaces $L^p (1 \leqslant p < \infty)$; also the theory of normed spaces and the theory of linear mappings. The principles of linear algebra are frequently used.

There is, for instance, the theorem of Weierstrass concerning the approximation of continuous functions by polynomials which I mentioned before. The generalization of this theorem is very instructive; it follows here.

Let X be a compact topological space. Denote by $C(X)$ the linear space of all continuous mappings of X into the field **R** of real numbers. $C(X)$ is a Banach-space if the norm of an element $f \in C(X)$ is defined by

$$\|f\| = \sup_{x \in X} |f(x)|.$$

But $C(X)$ is even an algebra, defining multiplication of the elements f, $g \in C(X)$ by

$$(fg)(x) = f(x) g(x).$$

---

[47]) A Paris, de l'imprimerie de la République. Prairial an V.

The generalization of the theorem of Weierstrass is as follows:

THEOREM OF STONE-WEIERSTRASS. *Let A be a subalgebra of C(X) which contains the constant functions and separates the points of X; then A is dense in the Banach-space* $C(X)$[48]).

The classical result of Weierstrass is a special case of this general theorem. The algebraic structure of $C(X)$ is an object of study (for instance the ideals of $C(X)$). More generally I mention the study of function algebras and, still more general, normed algebras where the elements of the algebra are no longer functions but elements of an abstract set.

In the domain of integration there is the theory of Haar-measure on groups, to which I will return below.

Normed spaces and algebras. A normed space E over $\mathbf{R}$ is a linear space over $\mathbf{R}$ together with a mapping $\|\cdot\| : E \rightarrow \mathbf{R}_+$ which satisfies

(i) $\|x\| = 0 \Leftrightarrow x = 0$;
(ii) $\|ax\| = |a| \cdot \|x\|$ for all $a \in \mathbf{R}$, $x \in E$,
(iii) $\|x + y\| \leqslant \|x\| + \|y\|$.

When the normed space E is complete, it is called a Banach-space. The normed space E is called a normed algebra, when a multiplication for the elements of E is defined such that

$$\|xy\| \leqslant \|x\| \cdot \|y\|.$$

In the modern theory of the integral, an integral is defined as a *linear functional* (linear form, linear function) defined on a function space, for instance the space $C(X)$.

The principle of such a definition of an integral is already to be found in the work of Lebesgue: in his descriptive definition his idea is to associate to a function a number, called the integral of the function, satisfying certain axioms (among which additivity). An integral is thus defined as a linear functional.

The development of the theory of functionals and functional analysis began already in the end of the nineteenth century. Although there are many relations between the history of the integral and the history of general analysis and functional analysis, I am not going to write a history of these last subjects. This should require a chapter for its own. I only give some indications which I think are necessary.

---

[48]) *A* separates the points of X if for any $x, y \in X$, $x \neq y$, there is a function $f \in A$ such that $f(x) \neq f(y)$.

JACQUES HADAMARD

The notion of a functional is due to Volterra. He considered real functions in which the variable is a curve or a function itself; this notion is related to functions of an infinite number of variables. In his work of 1887 [111] Volterra used the denomination "*fonction de ligne*". He speaks of quantities which depend on all the values of a function. The term "fonctionnelle" (functional) is due to HADAMARD. The object of functional analysis is the study of the properties of functionals. Historically this domain is thus more restricted than the general analysis of Moore and Fréchet. Nowadays, however, this general analysis is mostly called functional analysis, thus enlarging the classical meaning of this terminology. Volterra arrived at his study of functionals from the side of the calculus of variations with its well known classical problems[49]). He used the notation

$$F = F \left| \left[\varphi(x) \right] \right|_{a}^{b}.$$

The functionals he studied were not necessarily linear. He was led to linear functionals by the study of the variation of a functional which, when restricted to terms of the first order, is represented by an integral, that is a linear functional. Volterra applied his

---

[49]) I mention the Italian mathematician PINCHERLE and the French mathematician BOURLET who in the same time as Volterra were pioneers in this domain. For an extensive literature see [114].

theory of "fonctions de lignes" for instance in the theory of integral equations. For information on this theory see the books of Volterra [112], [113], [114]; see also Lévy [76].

I shall give the definition of the notion of linear functional, because it is most important for the modern definition of the integral.

Let E be a linear space over the field of the reals **R**. *A linear functional T is a mapping from E into* **R** *satisfying*

$$T(ax) = aT(x), \; a \in \mathbf{R}, \quad x \in E,$$
$$T(x+y) = T(x) + T(y), \quad x, y \in E.$$

When E is a topological linear space – for instance a normed space – it makes sense to speak of continuous linear functionals; continuity is defined in the well known way and, when E is a normed space, one proves that continuity is equivalent with boundedness of T, this means there is a constant $M \geqslant 0$ such that

$$|T(x)| \leqslant M \cdot \|x\|$$

for all $x \in E$.

For instance, the mapping

$$f \to {}_0\!\int^1 f(x)\,dx$$

defines a continuous linear functional on the normed space L^1 of the $\mathcal{L}$-integrable functions on $[0,1]$[50]).

JACQUES HADAMARD (1865-1963) was born in Versailles; he studied at the Ecole Normale Supérieure. In 1892 he wrote his thesis on the theory of functions which are defined by a Taylor-series. Having been Maître de Conférences in Bordeaux, he became Maître de Conférences at the Sorbonne in 1897 and professor at the Collège de France and the Ecole polytechnique in 1912. He worked in many different fields (functional analysis, partial differential equations).

FRÉDÉRIC RIESZ (1880-1956) was born in Györ (Hungary). He studied in Zürich, Budapest and Göttingen. After having finished his thesis in Budapest, he was for some time a teacher at secondary schools. Having been associate professor at the university of Kolozsvar, he became professor in Szeged in 1920 and in 1946 he went to Budapest. Riesz was one of the initiators of modern analysis; the development of functional analysis is to a great extent due to his investigations.

---

[50]) Some authors use the terminology linear form instead of linear functional.

FRÉDÉRIC RIESZ

In Szeged ALFRED HAAR, about whom we have to speak later, was one of Riesz's colleagues. Riesz and Haar founded the journal "Acta scientiarum Mathematicarum", which soon became a famous source of activity of the mathematical institute of the university of Szeged (the Bolyai-institute).

In 1903 Hadamard, in a note in the Comptes rendus de l'Académie des Sciences [53], posed the problem to find an analytical expression for the continuous linear functionals on the space $C(I)$ of the continuous functions on $[0,1] = I$. He answered this question in the following way. He obtained the result that, in the notation of Volterra, any linear continuous functional can be written in the form

$$T\,|\,[f(t)]\,| = \lim_{\mu \to \infty} {}_0\!\int^1 F(t,\mu)f(t)\,\mathrm{d}t,$$

where F is a continuous function.

This representation, however, is not unique.

In 1909 F. Riesz gave the definitive solution for this problem, using the integral defined by Stieltjes [96]. He proved that any continuous linear functional on $C(I)$ can be represented in the form of a Stieltjes-integral

$$T : f \to {}_0\!\int^1 f\,\mathrm{d}g,$$

where g is a function of bounded variation.

This theorem is known as the *Riesz representation theorem* (but there is an other one which also has this name, namely the representation of linear forms in a Hilbert-space).

This representation is canonical, i.e., the function g is uniquely determined if we agree to identify two functions of bounded variation g_1 and g_2, having the same points of continuity, when there is a constant C such that $g_1(x) - g_2(x) = C$ in all points of continuity (the set of the points of discontinuity of a function of bounded variation is at most enumerable). In that case the functions g_1 and g_2 generate the same Stieljes measure.

There is some reason to say that this theorem is the point at which the modern theory of the integral started. The Lebesgue-integral and the Stieltjes-integral are linear functionals. The procedure is now that, conversely, this property is taken as the definition of an integral for a certain class of functions. This leads to the most general definition of an integral, which has certain properties whose validity we think essential for the concept of an integral.

Such a method is common in modern mathematics: it is a way for building axiomatic theories. I mention for instance the theory of groups, the axiomatic theory of Hilbert space, the theory of normed spaces, the axiomatic potential theory. As will be clear the axiomatic method gives insight in the domain of validity of certain properties of mathematical objects.

On the basis of the work of Stieltjes, Fréchet, Moore, Radon this was done by YOUNG [117] and DANIELL [31].

I shall now give the definition of an integral along the lines it is defined by Bourbaki in his famous treatise on integration theory [18].

N. BOURBAKI is the name of a group of mathematicians, periodically renewing its members, which began to write about thirty years ago a series of books "Eléments de Mathématique" in which the leading principle is to place the structure of the mathematical theories in the foreground.

Let X be a locally compact topological space. Let $C(X)$ be the linear space of the real continuous functions on X with compact support[51]). Topologizing $C(X)$ by means of the norm

$$\|f\| = \sup_{x \in X} |f(x)|,$$

$C(X)$ is a normed space.

DEFINITION. *A measure (integral) μ is a linear functional on $C(X)$ such that for each compact $A \subset X$ there is a number $M_A \geqslant 0$ such that*

---

[51]) The support of a function f is the closure in X of the set of all $x \in X$ such that $f(x) \neq 0$.

$$|\mu(f)| \leqslant M_A \cdot \|f\|,$$

for every $f \in C(X)$ the support of which is contained in A.

The integral μ is said to be bounded if there is a constant $M \geqslant 0$ such that

$$|\mu(f)| \leqslant M \cdot \|f\|$$

for any $f \in C(X)$.

The value of μ for a function f is also denoted by

$$_X\!\!\int f \, d\mu \quad \text{or} \quad _X\!\!\int f(x) \, d\mu(x)$$

and is called the μ-integral of f. Some authors call μ a Radon-measure (or a measure).

In this way an integral is defined on a locally compact topological space X. When there is more structure in X, it is possible to require more properties for the integral. This leads for instance to the Haar-integral (see below). What is the relation between this general type of integral and the classical integrals? They are all included in this general definition for appropriate choice of X and μ: the Riemann-integral, the Lebesgue-integral, the Stieltjes-integral, all for continuous functions with compact support.

Example. Let (x_i) be a sequence of points in X; let (α_i) be a sequence of positive real numbers such that $\Sigma \alpha_i$ is convergent. Then

$$\mu(f) = \sum \alpha_i f(x_i) \quad (f \in C(X))$$

defines a measure on X. It is called a *discrete measure*.

Now, let a measure μ be given; the integral is defined for functions of $C(X)$. The problem is then to extend the domain of the definition of μ to a larger class than the continuous functions. This can be done by defining, with the aid of μ, an upper integral for any function on X (the upper integral may take the value $+\infty$). The upper integral is used to define a new norm in the linear space of all functions for which the upper integral is finite. The desired extension is then obtained by taking the closure of $C(X)$ in this space with respect to this new norm. This closure is the linear space L^1 of the μ-integrable functions.

As a special case, choosing μ in the right way, one obtains the space of the functions which are integrable in the sense of Lebesgue. But the method is essentially more general: one defines in this way a *family of integrals*.

The integral of a function is here primary; the measure of a μ-measurable set in X is obtained by taking the value of the μ-integral for the characteristic funtion of the set. Note that this method furnishes an integral on a locally compact space. In particular, starting from the Riemann-integral for continuous functions with compact support, it can be shown that one obtains the functions which are integrable in the sense of Lebesgue and the corresponding Lebesgue-integral on the real line $\mathbf{R}$, this means that in this case the integral is the same as the integral which we treated before with the analytical method of Lebesgue. The so called *improper* Riemann-integrals on **R**, obtained by a limiting process

$$_0\!\int^\infty f(x) = \lim_{a \to \infty} \; _0\!\int^a f(x)\,dx,$$

are not included in this procedure. As we remarked earlier, this method does not lead to an integral with good properties.

The theory of this general integral is then developed: sets of μ-measure zero (compare almost everywhere), an analogous notion for functions (negligible functions; classes of equivalent functions), convergence theorems for sequences of functions (f_n), the space L^p and so on. But now there are more properties than in the classical theory of the Lebesgue-integral.

The set of all linear forms on $C(X)$ of the kind we considered form a linear space, the so called *dual space* of $C(X)$. A topology can be defined on this space, called a weak topology in the theory of linear topological spaces (in integration theory it is sometimes called the vague topology). It then makes sense to speak of the convergence of a sequence (μ_n) of integrals. In this topology the integrals μ_n converge to the integral μ if

$$\lim_{n \to \infty} \; \mu_n(f) = \mu(f)$$

for every $f \in C(X)$.

Integration theory is thus connected with the theory of duality in the general theory of linear topological spaces.

3.2 *Generalizations*

The further development of modern integration theory has several aspects. Most of these are generalizations or improvements of the theory in the preceding paragraph. I give some indications, without claiming completeness.

One of the aims of the generalizations is to define an integral for functions which are no longer real valued.

1.   There is a theory of integration for functions which take their values in a Banach space E over the reals. Names of mathematicians who introduced such integrals are Bochner [7] and Pettis [92]. The value of the integral is not a real number but is itself also an element of a Banach space; this theory is connected with the theory of duality for Banach spaces. Functional-analytical methods are used. In absence of an order relation in E, all questions of convergence are to be treated with the aid of the norm in E.

For the definition and properties of these integrals we refer the reader to the book of Hille and Phillips [60], which contains also more bibliography (Birkhoff, Dunford, Gelfand). See also the treatise of Bourbaki.

2.   In the preceding generalization the space E is a linear space over the real number field. With respect to integration other fields are also studied. Especially there is a theory of integration of functions, defined on a topological space, which take their values in a non-archimedean valued field (the case of archimedean valued fields is not interesting in this respect, because it leads, owing to a theorem of Ostrowski, to the field of real numbers), for instance the p-adic number field. There is no order relation on such a field K and therefore the theory, developed analogously to the theory mentioned before (Bourbaki), is entirely based on the norm in K. Again, the value of the integral is not a real number, but an element of K. Some results of the classical (real) theory are valid for this case, but there are also differences (for instance the existence of a maximum set of zero measure). See Monna and Springer [83].

3.   A common feature of these two cases is that there can be no question of approximation of the integral from below and from above as is done for the Riemann-integral (Darboux-sums), the Lebesgue-integral and their generalizations (Daniell), the Perron-integral. Questions based on order are to be reduced to questions on the norm.

There is another theory of integration which is, on the contrary, exclusively based on order relations.

Consider, for a moment, an integral μ as before. It is said to be *positive* if $f \leqslant g$ implies $\mu(f) \leqslant \mu(g)$ where $f \leqslant g$ is meant in the sense of the natural ordering, that is $f(x) \leqslant g(x)$ for all $x \in X$. With regard to this property, a positive integral μ can be described as a mapping from the ordered set $C(X)$ into the ordered set of the real numbers which leaves the order relation invariant. This property can be taken as the starting point for the definition of an abstract integral. Now, remembering that one of the principles was to extend the domain of an integral, defined on a certain set of functions, to a larger set of functions, MCSHANE [80] formulates the following problem.

Let there be given two partially ordered sets F and G and an order-preserving mapping I_0 of a subset E of F into G. One asks for conditions on E, I_0, F and G which enables one to extend the domain of definition of I_0 in such a way that one gets a mapping I of a set F_0, containing E, into G which has certain useful continuity properties.

For a treatment of this problem we refer the reader to the book of McShane. The theory of Daniell is a special case of this general problem.

I mention two fundamental papers of Freudenthal [48], [49] in this direction.

4.   Various other generalizations and modifications in the theory of integration were

studied by several mathematicians. I can only mention some of them by way of example.

There are many generalizations in the literature of the Riesz-representation theorem, for instance with regard to spaces of functions which are not real valued, but which take their values in a locally convex space; a bibliography is in a paper of GOODRICE [51].

I gave a definition of a measure (integral) on a locally compact topological space. There are definitions of measure on topological spaces in which the condition of local compactness is weakened. This is a relatively recent theory for which the names of the mathematicians WIENER and PROKHOROV (1956) must be called. Convergence properties of measures can be proved in such cases. The theory is very important for the theory of probability. For such an integration theory I refer the reader to a paper of SONDERMANN [101] and to the chapter "Intégration sur les espaces topologiques séparés" in Bourbaki's Intégration [18].

Looman [78] studied integrals of the Denjoy type for functions of more real variables.

Integrals of the Denjoy type for functions – real valued or taking their values in a Banach space – whose domain is more general than the real line were studied by ROMANOVSKI and SOLOMON; for an account of these theories see [100].
Integrals of the Perron-type based on a generalized definition of differentiability were investigated. I mention for instance the *Cesaro-derivative*, where the increment of the function is replaced by the arithmetic mean of its increment. Putting

$$C(f,a,b) = \frac{1}{b-a}\int_a^b f(x)\,\mathrm{d}x,$$

the *Cesaro-upper-derivative* is then defined by

$$\overline{\lim_{h\to 0}}\ \frac{C(f,x,x+h)\ f-(x)}{\frac{1}{2}h};$$

the Cesaro-lower-derivative is defined as the limes inferior.
The integral of the Perron-type, which corresponds to these derivatives, is called the *Cesaro-Perron-integral*. See Burkill [23] and Kubota [67].

I mention the *upper* and *lower Burkill* integral of a function of an interval over an figure R_0, obtained as the upper resp. the lower limit of the sums of the values of the function corresponding to subdivisions of R_0. When these integrals are equal, their common value is called the *Burkill-integral* of the function. It is used in the theory of the area of a surface which we mentioned before. See [98], [65].

All these modifications of the integral have in common that they generalize or change in some way certain aspects of the classical definitions, such as the domain and the range of the function, the definition of differentiability.

CONSTANTIN CARATHÉODORY

3.3 *Carathéodory*

CONSTANTIN CARATHÉODORY (1873-1950) was born in Berlin, descending from an old Greek family. After a military training in Belgium and further study in Paris and London, he went as an engineer to Egypt. He turned, however, to mathematics and studied in Berlin and Göttingen. Having been professor in Hannover and Breslau, he went to Göttingen in 1903, succeeding Felix Klein. In 1918 he became professor in Berlin. After two years he founded in charge of the Greek government a Greek university in Smyrna (which was later destroyed). After two years in Athens, he settled as professor in München. Carathéodory published on many domains: analysis, thermodynamics, mechanics. He is especially known for his contributions to the theory of real functions and to measure theory and although we don't treat the theory of measure in our survey, it seems that some indications on his work cannot be omitted.

1.   I mentioned before the hierarchy of the sets which are Borel-measurable, the so called Borel-sets. The idea of their definitions was constructive: they are defined by means of transfinite application of some operations. It is clear that in this way it is difficult to get insight in the structure of these sets. This is the reason why an axiomatic definition of measurable sets was introduced; I already mentioned the notion of a σ-ring of sets.

Carathéodory has done important work on abstract measure theory. He gives a method for generating measures.

Let a metrical space M be given. Let Γ be a real function, defined on the subsets X of M; suppose $\Gamma(X) \geqslant 0$ for any $X \subset M$. Then Γ is called an outer measure in the sense of Carathéodory when the following conditions are satisfied:

(i) $\Gamma(X) \leqslant \Gamma(Y)$ if $X \subset Y$,

(ii) $\Gamma(\cup_i X_i) \leqslant \sum_i \Gamma(X_i)$ for any sequence $(X_i)_{i \in N}$

(iii) $\Gamma(X \cup Y) = \Gamma(X) + \Gamma(Y)$ if $\rho(X,Y) > 0$, ρ being the distance of the sets X and Y[52]).

The problem is, given an outer measure Γ, to define a notion of measurability with respect to Γ. Carathéodory gives the following definition.

The set $E \subset M$ is called Γ-measurable if

$$\Gamma(X) = \Gamma(X \cap E) + \Gamma(X \cap CE)$$

for every $X \subset M$[53]).

This seems a curious definition, but it was shown to be a very good one, because it had important consequences. Denote by $\mathfrak{L}_\Gamma$ the class of all Γ-measurable sets. The main problem is to prove that Γ is a measure on $\mathfrak{L}_\Gamma$ in the ordinary sense. The following facts hold:

(i)   *Γ is additive on $\mathfrak{L}_\Gamma$ and possesses the customary properties of a measure.*

(ii) *$\mathfrak{L}_\Gamma$ contains, for any Γ, the Borel-sets.*

Specializing Γ and M, one finds the classical measures. For the theory of Carathéodory and the corresponding integration theory we refer the reader to Carathéodory [26] and Saks [98].

2.   There is another direction in which Carathéodory has done interesting work on the theory of the integral [27].

In this work it is the aim of Carathéodory to produce a synthesis between the classical method of defining area by finite decompositions – a method which we described in section 1 – and the modern measure theory.

---

[52] d being the distance in M, the distance of X and Y is defined by
$$\rho(X,Y) = \inf d(x,y), \qquad x \in X, y \in Y.$$

[53]   I denote by CE the complement of E in M.

He remarks that this method is based on principles which are essentially different from those of the set-theoretic method. Indeed, when, in the decomposition method, a figure is decomposed in disjoint figures, the geometrical figures cannot be considered as point sets. What should be done with the "boundaries" of the figures? For instance for a triangle, are we to consider only interior points or must the sides be considered to belong to the triangle? For a decomposition in disjoint parts, neither is evidently possible. Nevertheless, the results of the decomposition method and the set-theoretic method agree with each other. This can be proved by considering the measure of a set and its closure, but Carathéodory wants to avoid this and he was thus led to a general theory of which both theories are a special case. This is done in an axiomatic way by introducing abstract objects, called *soma, somen*. These somen must have properties which on the one hand correspond to the set-theoretic properties of the subsets of a set and on the other hand correspond to the properties of the figures in elementary geometry. Carathéodory defines certain operations for somen, which are to be compared with the ring operations in algebra. Then a notion of function is defined which in the theory of the somen takes the place of the point functions in abstract spaces. This leads to somen functions, which are the base for the desired measure theory. Here the theory of area and measure has got an entirely algebraic character.

3.4 *Haar-measure*

Very important is the so called Haar-measure and the Haar-integral.

I remarked several times that the course of the development of integration theory led to consider functions, defined not on $\mathbf{R}^n$, but on an abstract set, if necessary provided with a topology or another structure.

Now, observe that the Lebesgue-measure has certain invariance properties: the measure of a set is invariant under translations. This is a consequence of the structure of the set of real numbers as an additive group.

This leads to a theory of integration of real functions, defined on a group, requiring certain conditions for the integral connected with the structure of the group. I give some definitions and results in this very important theory.

Let G be a topological group[54]). Let $C(G)$ be the space of the real-valued continuous functions on G with compact support. For any $y \in G$ and $f \in C(G)$, the function $fy \in C(G)$ is defined by

---

[54]) A topological group is a group on which a topology is defined in such a way that the group structure and the structure as a topological space are connected with each other by means of axioms, expressing continuity of the group operations.

ALFRED HAAR

$$(fy)(x)=f(yx), \quad x \in G.$$

Let μ be a measure (integral) as defined before (3.1). For any $y \in G$, define a measure $y\mu$ by

$$(y\mu)(f)=\mu(fy), \quad f \in C(X),$$

or

$$\int f \, \mathrm{d}(y\mu)=\int fy \, \mathrm{d}\mu.$$

DEFINITION. *The measure $\mu \neq 0$ is called a left invariant Haar-measure if $y\mu = \mu$ for every $y \in G$.*

The problem is to give conditions for the existence of such a measure. I give the main results, without entering into the history of them (the names of the mathematicians H. Poincaré, E. Cartan, A. Hurwitz are connected with this subject).

In 1933 A. Haar [52] proved the existence of a left invariant Haar-measure for any locally compact group having an enumerable base of open sets.

Speaking about Riesz, I already mentioned the mathematician Alfred Haar (1885-1933). Haar was first professor in Kolozsvár, later in Szeged.

In 1934 J. VON NEUMANN proved for compact groups that this measure is uniquely determined up to a constant factor [88].

Von Neumann and A. Weil generalized these theorems for general locally compact groups (Weil [115]).

Weil proved that the existence of an invariant measure on a group is characteristic for locally compact groups in so far that any Hausdorff topological group on which a left invariant measure exists is locally precompact.

Once these results where established, the way was open to the development of an analysis on locally compact groups, containing the classical analysis on **R** or **R**^{n} as a special case. This domain is still in full development; I refer the reader to the textbooks.

Generalizations of Haar-measure for non real-valued measures were also studied [83].

All this is again an illustration of the tendency towards algebraisation of analysis in modern mathematics.

I have confined myself in this survey to the strict history of the integral. Because this is not a textbook on the theory of integration, I have omitted the properties of the integral of which I mention only: product measures and the *theorem of Fubini*; the *theorem of Radon-Nikodym* which describes the relation between two integrals μ and v. I refer the reader to the textbooks and to the books on integration theory in the Bourbaki-series. I have also omitted information on the applications of modern integration such as: the application of vector-valued integrals to the recent theory of Choquet on the representation of the points of a convex set by means of vector-valued integrals, related to the famous theorem of Krein-Milman; a form of the Stieltjes-integral for representing certain linear operators in Hilbert-space (spectral analysis); abstract harmonic analysis, even in non-archimedean valued fields.

It should require another chapter to write about these things.

Bibliography

[1] BAILLAUD, B. and H. BOURGET. *Correspondance d'Hermite et de Stieltjes* I, II. Paris (1905).

[2] BAIRE, R. Sur les fonctions de variables réelles. *Annali di Mathematica* Série IIIa, t. III (1899).

[3] BAIRE, R. *Leçons sur les fonctions discontinues.* Paris (1905).

[4] BANACH, S. Sur le problème de la mesure. *Fundamenta Math. 4* (1923) pp. 7-33.

[5] BANACH, S. *Théorie des opérations linéaires.* Warszawa (1932).

[6] BANACH, S. and C. KURATOWSKI. Sur une généralisation du problème de la mesure. *Fundamenta Math. 14* (1929) pp. 127-131.

[7] BOCHNER, S. Integration von Funktionen, deren Werte die Elemente eines Vectorraumes sind. *Fundamenta Mat. 20* (1933) pp. 262-276.

[8] BOCKSTAELE, P. *Het intuïtionisme bij de Franse wiskundigen.* Verhandelingen van de Kon. Vlaamse Academie voor Wetenschappen, Letteren en Schone Kunsten van België. Brussel (1949)

[9] BOREL, E. *Leçons sur la théorie des functions.* Paris (1898).

[10] BOREL, E. *Leçons sur les fonctions entières.* Paris (1900).

[11] BOREL, E. *Leçons sur les séries divergentes.* Paris (1901).

[12] BOREL, E. *Leçons sur les fonctions de variables réelles et les développements en série de polynomes.* Paris (1905).

[13] BOREL, E. Le calcul des intégrales définies. *Journal de Math. Pures et Appl.* Sixième série, t. 8 (1912) pp. 159-210.

[14] BOREL, E. Sur l'intégration des fonctions non bornées et sur les définitions constructives. *Ann. Ecole Norm. Sup. (3), 36* (1919) pp. 71-92.

[15] BOREL, E. *Méthodes et problèmes de théorie des fonctions.* Paris (1922).

[16] BOULIGAND, G. *Les définitions modernes de la dimension.* Paris (1935).

[17] BOURBAKI, N. *Topologie générale* Ch. IX (1948) p. 25.

[18] BOURBAKI, N. *Intégration* chap. 1, 2, 3, 4; *Intégration des mesures*, chap. 5; *Intégration vectorielle*, chap. 6; *Mesure de Haar*, chap 7; *Convolution et représentations*, chap. 8; *Intégration sur les espaces topologiques séparés*, chap. 9; Hermann, Paris.

[19] BOUTROUX, P. Sur la notion de correspondance. *Revue de Métaphysique et de Morale 12* (1904) pp. 909-920.

[20] BOUTROUX, P. *L'idéal scientifique des mathématiciens dans l'Antiquité et dans les temps modernes*. Paris (1920).

[21] BOYER, C. B. Cavalieri, limits and discarded infinitesimals. *Scripta Math. 8* (1941) pp. 79-91.

[22] BOYER, C. B. Proportion, equation, function: three steps in the development of a concept. *Scripta Math. 12* (1946) pp. 5-13.

[23] BURKILL, J. C. The Cesaro-Perron integral. *Proc. London Math. Soc. (2), 34* (1932) pp. 314-322.

[24] BURKILL, J. C. Henri Lebesgue. *J. London Math. Soc. 19* (1944) pp. 56-64.

[25] CANTOR, G. De la puissance des ensembles parfaits de points. *Acta Math. 4* (1884) pp. 381-392.

[26] CARATHÉODORY, C. *Vorlesungen über reelle Funktionen*. Leipzig-Berlin (1918).

[27] CARATHÉODORY, C. *Mass und Integral und ihre Algebraisierung*. Basel (1956).

[28] CAUCHY, A.-L. Mémoire sur les fonctions continues. *C. R. Acad. Sc. Paris XVIII* (1844) p. 116.

[29] CAUCHY, A.-L. *Œuvres complètes*, le série, t. VIII.

[29a] CESARI, L. Surface area. *Annals of Mathematics Studies 35*, Princeton (1956).

[30] COLLINGWOOD, E. F. Emile Borel. *Journal London Math. Soc. 34* (1959) pp. 488-512.

[31] DANIELL, P. J. A general form of integral. *Ann. of Math. (2) t. XIX* (1918) pp. 279-294.

[31a] DARBOUX, G. Mémoire sur les fonctions discontinues. *Ann. Ecole Norm. Sup. (2) IV* (1875) pp. 57-112.

[32] DAVID, L. V. Die beiden Bolyai. *Beiheft no. 11 zur Zeitschrift "Elemente der Mathematik"*. Basel (1951).

[33] DEHN, M. Über den Rauminhalt. *Math. Ann. 55* (1901) pp. 465-478.

[34] DENJOY, A. Mémoire sur les nombres dérivés des fonctions continues. *Journ. de Math. Pures et Appl. (7) 1* (1915) pp. 105-240.

[35] DENJOY, A. Sur les fonctions dérivées sommables. *Bull. Soc. Math. France 43* (1915) pp. 161-248.

[36] DENJOY, A. Mémoire sur la totalisation des nombres dérivés non-sommables. *Ann. Ecole Norm. Sup. 33* (1916) pp. 127-222.

[37] DENJOY, A. *Jubilé scientifique* (1955).

[38] DENJOY, A., L. FÉLIX and P. MONTEL. Henri Lebesgue le savant, le professeur, l'homme, *L'Enseignement Mathématique série II t. III* (1957) pp. 1-18.

[39] DIJKSTERHUIS, E. J. *De elementen van Euclides* I, II. Groningen (1929).

[40] DIJKSTERHUIS, E. J. *Archimedes.* Kopenhagen (1956).

[41] ENRIQUES, F. *Fragen der Elementargeometrie* I. Leipzig, Berlin (1911).

[41a] FEDERER, A. *Geometric measure theory.* Berlin (1969).

[42] FLECKENSTEIN, J. G. Der Prioritätsstreit zwischen Leibniz und Newton. *Beiheft no. 12 zur Zeitschrift "Elemente der Mathematik"* Basel (1956).

[43] FRÉCHET, M. Sur quelques points du calcul fonctionnel. *Rend. Palermo, t. XXII* (1906) pp. 1-74.

[44] FRÉCHET, M. Sur l'intégrale d'une fonctionnelle étendue à un ensemble abstrait. *Bull. Soc. Math. France 43* (1915) pp. 248-265.

[45] FRÉCHET, M. *Les espaces abstraits et leur théorie considérée comme introduction à l'analyse générale* (collection Borel). Paris (1928).

[46] FRÉCHET, M. *Pages choisies d'analyse générale,* Paris (1953).

[47] FRÉCHET, M. *La vie et l'œuvre d'Emile Borel.* Monographies de l'Enseignement Mathématique no. 14. Genève (1965).

[48] FREUDENTHAL, H. Teilweise geordnete Moduln. *Proc. Kon. Ned. Akad. v. Wetensch. 39* (1936) pp. 641-651.

[49] FREUDENTHAL, H. Zur Abstraktion des Integralbegriffs. *Proc. Kon. Ned. Akad. v. Wetensch. 39* (1936) pp. 741-745.

[50] GAUSS, C. F. *Werke B8.*

[51] GOODRICH, R. K. A Riesz representation theorem. *Proc. Am. Math. Soc 24* (1970) pp. 629-636.

[52] HAAR, A. Der Massbegriff in der Theorie der Kontinuierlichen Gruppen. *Ann. of Math. (2) t. XXXIV* (1933) pp. 147-169.

[53] HADAMARD, J. *Sur les opérations fonctionnelles.* C. R. Acad. Sc. Paris (1903).

[54] HADWIGER, H. Ergänzungsgleichheit k-dimensionaler Polyeder. *Math. Z. 55* (1952) pp. 292-298.

[55] HADWIGER, H. *Vorlesungen über Inhalt, Oberfläche und Isoperimetrie.* Springer-Verlag (1957).

[56] HAUSDORFF, F. *Grundzüge der Mengenlehre.* Leipzig (1914).

[57] HAUSDORFF, F. *Mengenlehre,* third edition Dover publ.

[58] HILBERT, D. *Grundlagen der Geometrie.*

[59] HILBERT, D. *Gesammelte Abhandlungen.* Bd. III.

[60] HILLE, E. and R. S. PHILLIPS. Functional analysis and semi-groups. *A.M.S.* New York (1957).

[61] JESSEN, B. The algebra of polyhedra and the Dehn-Sydler Theorem. *Math. Scard. 22,* (1968) pp. 241-256.

[62] JORDAN, C. *Cours d'analyse de l'école polytechnique.* 2ème édition Paris (1893).

[63] KAMKE, E. *Das Lebesguesche Integral.* Berlin-Leipzig (1925).

[64] KAMKE, E. *Das Lebesgue-Stieltjes Integral.* Leipzig (1956).

[65] KEMPISTY, S. *Fonctions d'intervalle non additives.* Paris (1939).

[66] KRONECKER, L. *G. Lejeune Dirichlets Werke* Bd. I (1889) p. 133.

[67] KUBOTA, Y. On a characterization of the C.P.-integral. *Journ. London Math. Soc. 43*, (1968) pp. 607-611.

[68] LAGRANGE, J. L. *Théorie des fonctions analytiques.* Paris (1797).

[69] LEBESGUE, H. Intégrale, Longueur, Aire. *Annali di Mathematica, Serie III, t. VII* (1902) pp. 231-359.

[70] LEBESGUE, H. *Leçons sur l'intégration et la recherche des fonctions primitives.* Paris (1903) 2nd ed. (1928).

[71] LEBESGUE, H. Sur les fonctions représentables analytiquement. *Journ. de Math. Pures et Appl. (6) 1* (1905) pp. 139-216.

[72] LEBESGUE, H. Remarques sur les théories de la mesure et de l'intégration. *Ann. Ecole Normale Sup. (3) 35* (1918) pp. 191-250.

[73] LEBESGUE, H. Sur l'équivalence des polyèdres réguliers. *C. R. Acad. Sc. Paris 907* (1938) pp. 437-439.

[74] LEBESGUE, H. Une fonction continue sans dérivée. *l'Enseignement Math. t. 38* (1942) pp. 212-213.

[75] LEBESGUE, H. *Notices d'histoire des mathématiques.* Monographies de l'Enseignement Mathématique no. 4. Genève (1958).

[76] LÉVY, P. *Leçons d'analyse fonctionnelle.* Paris (1922).

[77] LÉVY, P., S. MANDELBROJT, B. MALGRANGE and P. MALLIAVIN. *La vie et l'œuvre de Jaques Hadamard.* Monographie no. 16 de l'Enseignement Mathématique. Genève (1967).

[78] LOOMAN, H. Sur la totalisation des dérivées des fonctions continues de plusieurs variables indépendantes. *Fundamenta Math. 4* (1923) pp. 246-285.

[78a] LUSIN, N. Sur une question concernant la propriété de Baire. *Fundamenta Math. IX* (1927) pp. 116-118.

[79] LUXEMBURG, W. A. J. *Non standard analysis.* California Inst. of Techn., Pasadena, California (1962).

[80] McSHANE, E. J. *Order-preserving maps and integration processes.* Princeton (1953).

[81] MONNA, A. F. Sur le problème de la mesure. *Proc. Kon. Ned. Akad. v. Wetensch. 49* (1946) pp. 63-64.

[82] MONNA, A. F. Problème des moments et fonctions quasi-analytiques. *Nieuw Archief voor Wiskunde* 3e serie XVII (1969) pp. 189-199.

[83] MONNA, A. F. and T. A. SPRINGER. Intégration non-archimédienne I, II. *Proc. Kon. Ned. Akad. v. Wetensch. A66* (1963) pp. 634-642, pp. 643-653.

[84] MONTEL, P. and A. ROSENTHAL. Integration und Differentiation. *Enzyklopädie der Math. Wiss.* 2, 3, II (1923) p. 1044.

[85] MOORE, E. H. On the theory of systems of integral equations of the second kind. *Bull. Am. Math. Soc. 12, 280* (1906) pp. 283-284.

[86] MOORE, E. H. *General Analysis* I, II. Philadelphia (1935).

[87] NAGATA, J. I. *Modern dimension theory.* Groningen (1965).

[88] NEUMANN, J. VON. Zum Haarschen Mass in topologischen Gruppen. *Comp. Math. I* (1934) pp. 106-114.

[89] PAPLAUSKAS, A. B. L'influence de la théorie des séries trigonométriques sur le développement du calcul intégral. *Archives intern. d'histoire des Sciences 21, no. 84-85* (1968) pp. 249-260.

[90] PEANO, G. *Applicazioni geometriche del calculo infinitesimale.* Turin (1887).

[91] PERRON, O. Über den Integralbegriff. *S. B. Heidelberg Akad. Wiss. 16* (1914).

[92] PETTIS, B. J. On integration in vector spaces. *Trans. Amer. Math. Soc. 44* (1938) pp. 277-304.

[93] RADO, T. *On the problem of Plateau.* Berlin (1933).

[94] RADON, J. Theorie und Anwendungen der absolut additiven Mengenfunktionen. *S. B. der Math. naturwiss. Klasse der Akad. der Wiss. Wien t. CXXII, Abt. IIa* (1913) pp. 1295-1438.

[95] RIEMANN, B. *Gesammelte Mathematische Werke und Wissenschaftlicher Nachlass* herausgegeben unter Mitwerkung von Richard Dedekind von Heinrich Weber. Dover publications Inc., New York N.Y.

[96] RIESZ, F. *Sur les opérations fonctionnelles linéaires.* C. R. Acad. Sc. Paris (1909).

[97] ROBINSON, A. *Non-standard analysis.* Amsterdam (1966).

[98] SAKS, S. *Theory of the integral.* Warszawa (1937).

[99] SCHLESINGER, L. and A. PLESSNER. *Lebesguesche Integrale und Fouriersche Reihen.* Berlin, Leipzig (1926).

[100] SOLOMON, D. W. Denjoy integration in abstract spaces. *Memoirs of the Amer. Math. Soc. no. 85* (1969).

[101] SONDERMANN, D. Masse auf lokalbeschränkten Räumen. *Ann Inst. Fourier 19, 2* (1969) pp. 33-113.

[102] STIELTJES, TH. J. Recherches sur les fractions continues. *Ann. Sci. Toulouse (1) 8* (1894) J 1-122, *(1) 9* (1895) A 1-47.

[103] STRUIK, D. J. *A concise history of mathematics*, London (1962).

[104] SYDLER, J.-P. Sur la décomposition des polyèdres. *Comment. Math. Helv. 16* (1943/44) pp. 266-273.

[105] SYDLER, J.-P. Conditions nécessaires et suffisantes pour l'équivalence des polyèdres de l'espace euclidien à trois dimensions. *Comment. Math. Helv. 40* (1965/66) pp. 43-80.

[106] TARSKI, A. Une contribution à la théorie de la mesure. *Fundamenta Math. 15* (1930) pp. 42-50.

[107] TROPFKE, J. *Geschichte der Elementar-Mathematik.* 2e Auflage, Berlin/Leipzig (1921).

[108] VALLÉE POUSSIN, C. DE LA. *Intégrales de Lebesgue, fonctions d'ensemble, classes de Baire.* Paris (1916).

[109] VITALI, G. *Sul problema della misura dei gruppi di punti di una retta.* Bologne (1905).

[110] VOLTERRA, V. Sui principii del calcolo integrale. *Giorn. Mat. Battaglini 19* (1881) pp. 333-372.

[111] VOLTERRA, V. Sopra le funzioni che dipendono da altre funzioni. *Rend. Ac. Lincei* série 4, t. 3 (1887).

[116] VOLTERRA, V. *Leçons sur les fonctions de lignes.* Paris (1913).

[113] VOLTERRA, V. *Leçons sur les équations intégrales et les équations intégro-différentielles.* Paris (1913).

[114] VOLTERRA, V. and J. PÉRÈS. *Théorie générale des fonctionnelles.* Paris (1936).

[115] WEIL, A. L'intégration dans les groupes topologiques et ses applications. *Actual. Sc. et Ind. no. 869* Paris (1940, 1953).

[116] WEYL, H. Die heutige Erkenntnislage in der Mathematik. *Symposium 1* (1925) 1-32; *Gesammelte Abhandlungen II* 511-542.

[117] YOUNG, W. H. A new method in the theory of integration. *Proc. London Math. Soc. (2)* t. IX (1911) pp. 15-50.

[118] ZAANEN, A. C. *An introduction to the theory of integration.* Amsterdam (1958).

Added in proof:

[119] HAWKINS, TH. Lebesgue's theory of integration. The University of Wisconsin Press (1970).

Index

List of Mathematicians

Ackermann, Wilhelm	(1896-1962)
Archimedes	(287- 212 b.C.)
Baire, Réné	(1874-1932)
Banach, Stefan	(1892-1945)
Bernays, Paul	(1888-)
Bernoulli, Johann I.	(1667-1748)
Berstein, Felix	(1878-1956)
Bois-Reymond, Paul du	(1831-1889)
Bolyai, Johann	(1802-1860)
Bolyai, Wolfgang	(1775-1856)
Bolzano, Bernard	(1781-1848)
Borel, Emile	(1871-1956)
Brouwer, Luitzen E. J.	(1881-1966)
Burali-Forti, Cesaro	(1861-1931)
Cantor, Georg	(1845-1918)
Carathéodory, Constantin	(1873-1950)
Cauchy, Augustin	(1789-1857)
Cavalieri, Bonaventura	(1598-1647)
Cohen, Paul J.	(1934-)
Danieli. P. J.	(1889-1946)
Darboux, Gaston	(1842-1917)
Dedekind, Richard	(1831-1916)
De la Vallée Poussin, Charles	(1866-1962)
Denjoy, Arnaud	(1884-)
Descartes, René	(1596-1650)
Dirichlet, G. Lejeune	(1805-1859)
Euclides	about 300 b.C.
Eudoxos	(408- 355 b.C.)
Euler, Leonhard	(1707-1783)
Fermat, Pierre de	(1601-1665)
Fraenkel, Abraham A.	(1891-1965)
Fréchet, Maurice	(1878-)
Frege, Gottlob	(1848-1925)
Galilei, Galileo	(1564-1642)
Gauss, Carl Friedrich	(1777-1855)
Gödel, Kurt	(1906-)
Haar, Alfred	(1885-1933)
Hadamard, Jacques	(1856-1963)
Hartogs, Friedrich	(1874-1943)
Hausdorff, Felix	(1874-1942)
Hessenberg, Gerhard	(1874-1925)
Hilbert, David	(1862-1943)

161

Huygens, Christiaan	(1629-1690)	Radon, Johann	(1887-1956)
		Riemann, Bernhard	(1826-1866)
Jordan, Camille	(1838-1922)	Riesz, Frédéric	(1880-1956)
		Russell, Bertrand	(1872-1970)
Kepler, Johannes	(1571-1630)		
König, Julius	(1849-1913)	Saint Vincent, Grégoire de	(1584-1667)
Kronecker, Leopold	(1823-1891)	Schoenfliess, Arthur	(1853-1928)
		Scott, Dana	(1932-)
Lagrange, Louis de	(1736-1813)	Sierpinski, Waclaw	(1882-1969)
Lebesgue, Henri	(1875-1941)	Skolem, Thoralf	(1887-1963)
Leibniz, Gottfried Wilhelm	(1646-1716)	Stieltjes, Thomas Jan	(1856-1894)
Mirimanoff, Dimitri	(1861-1945)	Tarski, Alfred	(1901-)
Mittag Leffler, Gustav	(1846-1927)		
Moore, Eliakim Hastings	(1862-1932)	Vitali, Guiseppe	(1875-1932)
Mostowski, Andrzej	(1913-)	Volterra, Vito	(1860-1940)
Neumann, Johann von	(1903-1957)	Wallis, John	(1616-1703)
Newton, Isaac	(1642-1727)	Whitehead, Alfred North	(1861-1947)
		Wiener, Norbert	(1894-1964)
Pascal, Blaise	(1623-1662)	Wittgenstein, Ludwig	(1889-1951)
Peano, Guiseppe	(1858-1932)		
Perron, Oskar	(1880-)	Young, William Henri	(1863-1942)
Poincaré, Henri	(1854-1913)		
		Zermelo, Ernst	(1871-1956)
Quine, Willard van Orman	(1908-)		

162